16th Offshore Symposium

Energy from the Oceans

Houston, Texas, USA
9 February 2010

ISBN: 978-1-5108-3877-2

Printed from e-media with permission by:

Curran Associates, Inc.
57 Morehouse Lane
Red Hook, NY 12571

Some format issues inherent in the e-media version may also appear in this print version.

Printed by Curran Associates, Inc. (2017)

Society of Naval Architects and Marine Engineers (SNAME) Texas Section
99 Canal Center Plaza, Suite 310
Alexandria, VA 22314
USA

Phone: +1-703-997-6701

www.sname.org/texassection

Additional copies of this publication are available from:

Curran Associates, Inc.
57 Morehouse Lane
Red Hook, NY 12571 USA
Phone: 845-758-0400
Fax: 845-758-2633
Email: curran@proceedings.com
Web: www.proceedings.com

PROCEEDINGS
OF
THE 16TH OFFSHORE SYMPOSIUM

ENERGY FROM THE OCEANS

HOUSTON, TEXAS – FEBRUARY 9, 2010

sponsored by the

TEXAS SECTION
OF THE
SOCIETY OF NAVAL ARCHITECTS
AND
MARINE ENGINEERS

TABLE OF CONTENTS

(This page intentionally left blank.)

SESSION A PAPERS

(This page intentionally left blank.)

DEVELOPMENT OF WIND POWER IN THE GULF OF MEXICO

Rolando E. Vega, Ph.D., P.E.,
ABS Consulting

ABSTRACT

Several leases have been granted for the development of wind power in the Northern Gulf of Mexico. This paper reviews the wind resource potential in the Gulf of Mexico (GOM) and the need for meteorological data collection. Furthermore, a review of the current wind turbine technology is presented with an emphasis on suitability for hurricane-prone areas. An examination of the GOM bathymetry and current support structures needed to develop wind power in the region is presented. Hurricane risk in the GOM, with emphasis in ways to assess the wind and wave risk to offshore wind structures is introduced. A coupled metocean methodology is presented for extreme hurricane conditions. The need for wind turbine optimized designs is discussed with the challenge to use independent extreme hurricane events to determine design loads from extrapolation of load effects. Previous investigations sponsored by the Minerals Management Service (MMS) suggest that the higher variability (coefficients of variation) in the estimation of hurricane intensities results in lower structural reliabilities. The applicability of this statement for the US coast is evaluated.

1.0 INTRODUCTION

Despite the progress made in terms of regulations and proposed projects, no offshore wind project has been built in state or federal U.S. waters as of today. It is generally believed that offshore wind power will be first developed off the eastern coast of the U.S. where hurricane risk is lower and several leases have already been awarded. However, there remains an ongoing debate as to the exact location of the first offshore wind project given the complex regulatory process established at both the state and federal levels.

This presentation addresses the three main sources of uncertainty facing owners considering offshore wind development, specifically in the Gulf of Mexico. These primary concerns are focused on technical, environmental and financial risks. This paper highlights the heightened interest in the development of offshore wind power projects in the Gulf of Mexico (Schellstede, 2007) and the challenges associated with such development.

This presentation concludes by providing a brief background into the unique environmental loads and structural behavior as well as the probabilistic tools employed by industry to determine these loads as they relate to offshore wind structures including the turbines. This technical background will provide a foundation for those considering development and the risks associated with offshore wind development in the Gulf of Mexico.

2.0 WIND POWER POTENTIAL IN THE GOM

An early estimate of the wind energy potential in the Gulf of Mexico was performed by the Pacific Northwest Laboratory (PNL) as part of a worldwide wind resource assessment (Cherry et al., 1981). A subsequent study in 1987 by PNL (Elliot, 1987) presented a second generation more comprehensive evaluation of the wind resource in Central America and the Caribbean. During this second evaluation, new methodologies and data were developed including a detailed assessment of one-degree quadrangle ship data containing extensive geographic coverage and excellent exposure to winds from all directions. Results from this re-assessment in 1987 are illustrated by the average annual wind speeds (in meters per second) at 10 meters shown in Figure 1.

More refined modeling can now be performed to obtain wind resource estimates given the improvement in computational power and data collected in the last two decades. Today, with the use of comprehensive buoy, satellite, ship, coastal and offshore meteorological stations, numerical models are rendered to provide a geo-referenced and validated wind energy assessment. A coastline up to 50 nautical miles away from the coast has been re-assessed with these newer technologies in a grid size of 0.2 km by 0.2 km (Elliot and Schwartz, 2006). An example of this assessment for the Gulf of Mexico can be seen in the wind energy map produced by the National Renewable Energy Laboratory – excerpt shown in Figure 2.

There is good wind energy potential along the Gulf of Mexico coastlines, with excellent resources in the southern Texas coast. However, when compared to the outstanding wind resource in the eastern coast of the U.S., much of the wind power development attention has been withdrawn from the Gulf of Mexico and directed to the eastern coast where wind resources over land and severity of hurricane intensity are less. The shallowest coast with greatest wind resources are in the U.S. east coast.

Wind resources in north Texas are significant and has been the focus of wind industry development for the last decade or so. However, transmission challenges to carry wind energy production from the north to large load centers throughout the state have encouraged consideration of offshore wind power in the state. Current estimates by the U.S. Offshore Wind Collaborative (2009) suggest that the Gulf of Mexico has a wind energy potential with wind power Class 5 or

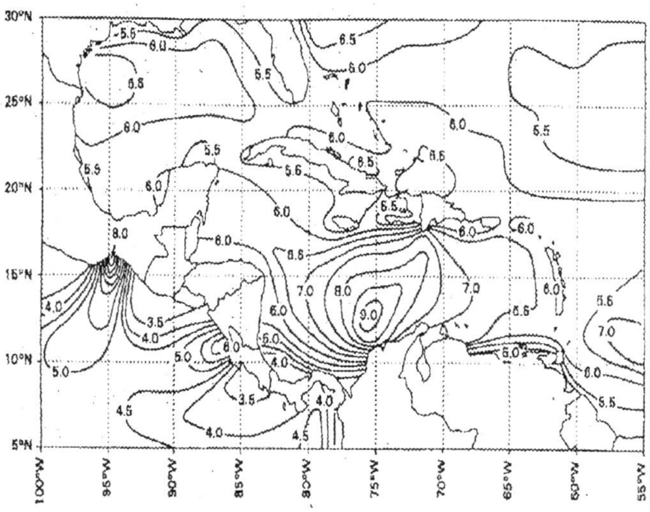

Figure 1: Annual average wind speed (m/s) [Elliot, 1987]

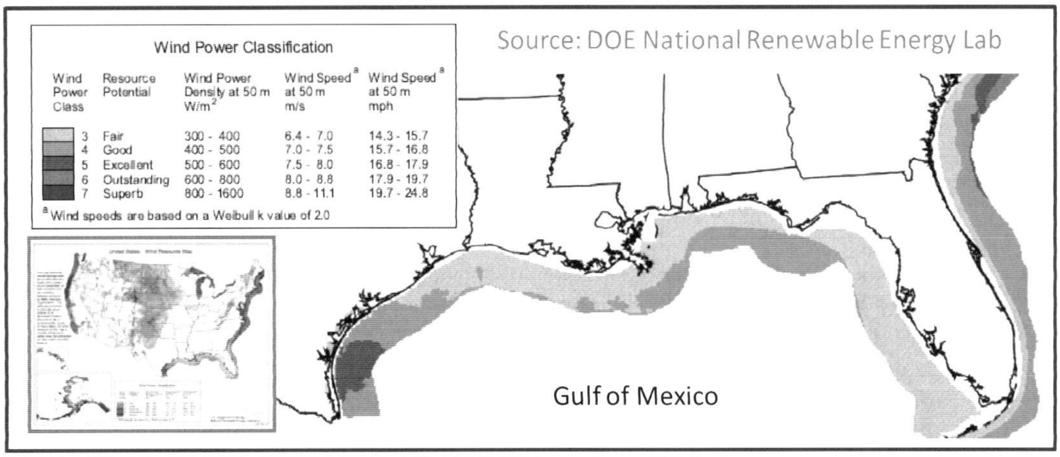

Figure 2: Wind power potential in the GOM

greater in excess of 67 GW.

In early stages, the onshore wind development in the U.S. focused in the windiest sites of Class 6 or above. With the development of improved, larger wind turbines and progress in construction methods, operations and maintenance, lower energy costs were possible thus becoming economically attractive to install wind turbines at sites with lower wind energy resource. Today, economically viable options may be justified onshore for sites with a wind power Class 3 or above. Similar progress will be necessary for offshore wind power development. Wind energy developers in the Gulf of Mexico can draw from the vast oil and gas industry experience to increase the number of opportunities and projects that are economically viable for sites with wind power Class 3 and above. This area covers virtually the entire Gulf.

Rigorous meteorological measurement campaigns are necessary for any wind project on- and offshore. In the GOM offshore environment, analysis with validated wind resource assessments and uncertainty analysis will be a critical step to establish an economically viable project.

3.0 CURRENT STATE OF OFFSHORE WIND TURBINE TECHNOLOGY FOR HURRICANE CONDITIONS

The vast majority of offshore wind turbine installations in Europe are extensions of land-based turbine technology that may not be optimized for operations at sea (Musial and Butterfield, 2006). Nonetheless, these wind turbines for offshore applications have been adapted to marine environments. Electrical and corrosion prevention system in the nacelle are retrofitted for these new conditions. While a 3-bladed wind turbine is the common technology and state of the practice today, 2-bladed wind turbines have technical and economical advantages to offshore wind. A list of commercial wind turbines with offshore capabilities, models available or under development are shown in Table 1.

According to a study by Penn State University (Stewart, 2009), most wind turbine designs (as specified by the three wind turbine classifications in IEC 61400) can only withstand up to Category 3 hurricanes in the Saffir Simpson Scale. In Stewart's study, the failure to consider height corrections for wind speed only makes the situation more critical. By considering the hurricane risk in the U.S. and the wind turbine classes specified in the International Electrotechnical Commission (IEC, 2005) it is found that the less conservative wind turbine design (i.e. IEC Class III) is appropriate for most of the U.S. onshore wind applications aside from complex terrain or special cases. On the other hand, after consideration of correction factors for averaging time and height, as specified in IEC 61400-1 for the most conservative wind turbine class (i.e. IEC Class I) at a height of 80 meters, no standard wind turbine would be able to survive a wind speed in excess of 125 mph (3-second at 10 meters over open terrain). In many instances, wind turbine design is governed by operational wind speeds (between cut-in and cut-out speeds) where the turbine is producing power and planned high thrust forces, not

Table 1: Commercial wind turbines with offshore models (Courtesy of Walt Musial, National Renewable Energy Laboratory)

Turbine Manufacturer	Model/Power	Date Available	Offshore operating experience
Siemens	SWT 2.0 (2MW)	2000	Commercial
Vestas	V80 2MW	2000	Commercial
General Electric (GE)	GE 3.6MW	2003	Commercial Inactive
Siemens	SWT 2.3 (2.3MW-107)	2003	Commercial
Vestas	V90 3MW	2004	Commercial
Siemens	STT 3.6 (3.6MW)	2005	Commercial
RePower Systems (Suzlon)	5M (5MW)	2005	45meter Demonstration/Commercial
Multibrid (Areva)	M5000 (5MW)	2005	Onshore 2005 Borkom West Pilot
Nordex	N90 (2.5MW)	2006	Offshore Demonstration 2003
RePower Systems (Suzlon)	6M (6MW)	2009	Prototype Assembled
Bard Engineering	VM (5MW)	2009	Onshore Prototype 2008
Siemens	SWT 3.6 (3.6MW-120)	2009	Prototype released 09/09
Vestas	V112 (3MW)	2009	Prototype released 09/09
Clipper Windpower	LIBERTY (2.5MW)	-	Commercial onshore
Clipper Windpower	BRITANNIA (10MW)	-	Prototype design stage

when the blades are feathered. A careful independent evaluation is in order in such cases as to demonstrate that the turbine is suitable for the location.

The U.S. national wind map shown in ASCE 7 (ASCE 7, 2006) defines the likely wind speeds for a 50 year event. IEC Class I wind turbines (the most conservative) should survive a strong Category 2 hurricane, as shown in Table 2. Most U.S. offshore and coastal locations in hurricane-prone areas with good wind resources will need a design of wind turbine known in IEC as "special design", or a careful independent evaluation of the wind turbine for the specific site. This is illustrated in Figure 3 with the ASCE 7 wind map. Offshore projects in the U.S. northeast coast may be suitable for use with a standard IEC wind turbine.

Table 2: IEC wind turbine classes and Saffir-Simpson hurricane classifications

Wind turbine class	Approximate 3-sec wind speed @ 10m in open terrain exposure (mph)	Saffir Simpson Scale	Application in the U.S.
IEC-I	125	Moderate to Strong Category 2 winds or less	Up to 125mph in ASCE 7 wind map
IEC-II	106	Moderate Category 1 Hurricane or less	Up to 106mph in ASCE 7 wind map
IEC-III	93	Weak Category 1 Hurricane	Non-hurricane areas of the US (most onshore wind)
IEC-S	-	-	Most U.S. Offshore and certain coastal areas

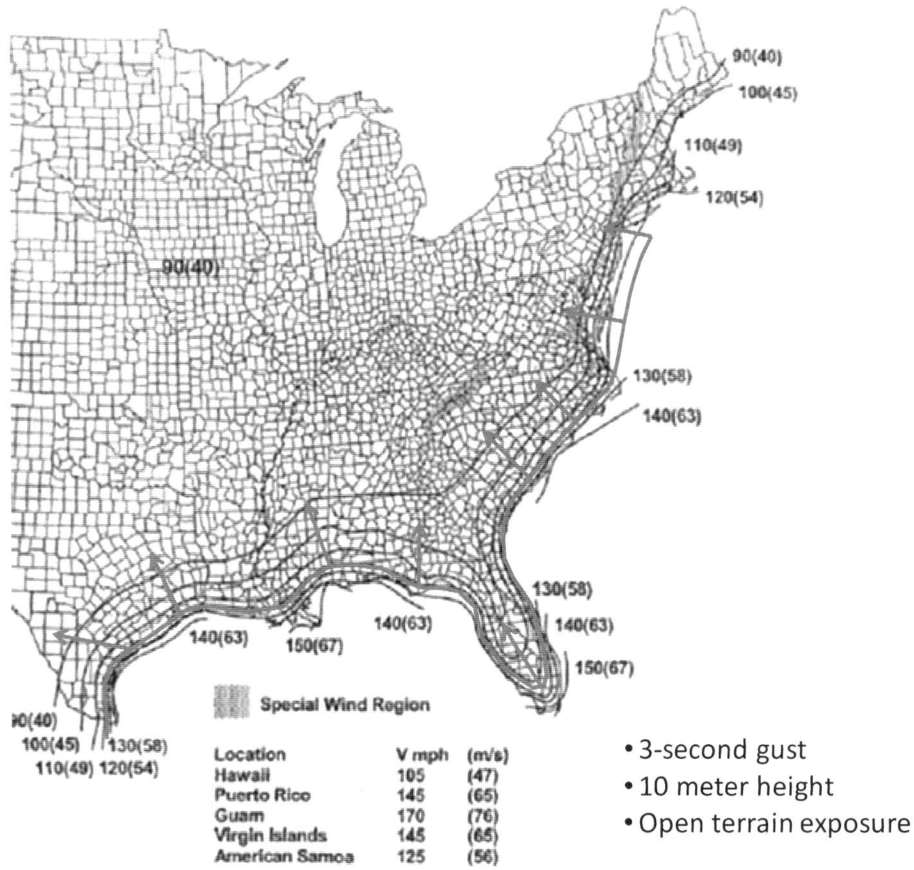

Figure 3: ASCE 7 (ASCE, 2006) wind map with approximate regions where Standard IEC wind turbines (Class I, II, & III) may be appropriate without special designs or further justification

4.0 BATHYMETRY IN THE GOM AND WIND TURBINE SUPPORT STRUCTURES

Another key consideration when assessing the potential and economics of offshore wind farm development in the offshore environment is the water depth. This is a driving parameter to establish the technology and initial capital cost required for the project. The latest generation of publicly available wind resource assessments covers 50 nautical miles from the shore. This is more or less the spatial coverage for approximately 20-30 meters of water depth. Approximately 38% of the Gulf is comprised of shallow and intertidal areas (< 20 m deep) (Gore,

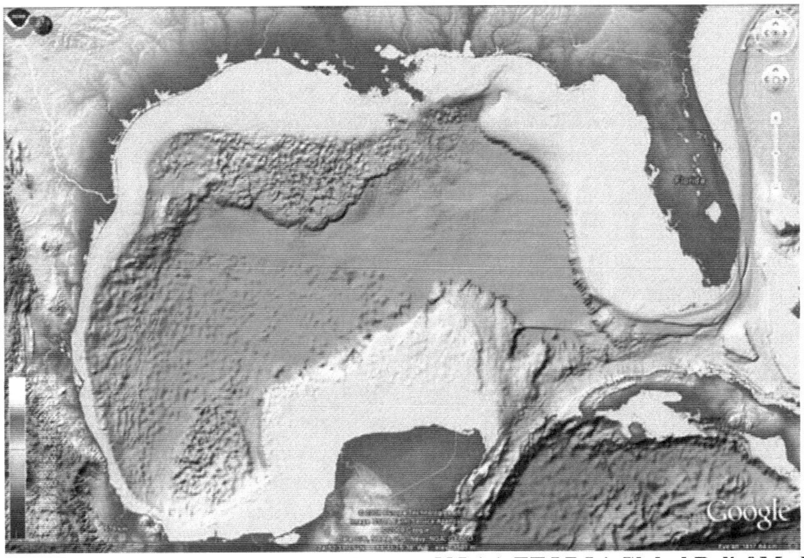

Figure 4: Bathymetry of the GOM [Source: NOAA ETOPO1 Global Relief Model]

Table 3: Depth in the GOM per region

GOM region	Water depth [m]	GOM representation
Shallow and intertidal areas	0-20m	38%
Continental shelf	20-180m	22%
Continental slope	180-3000m	20%
Abyssal areas	3000-4000m	20%

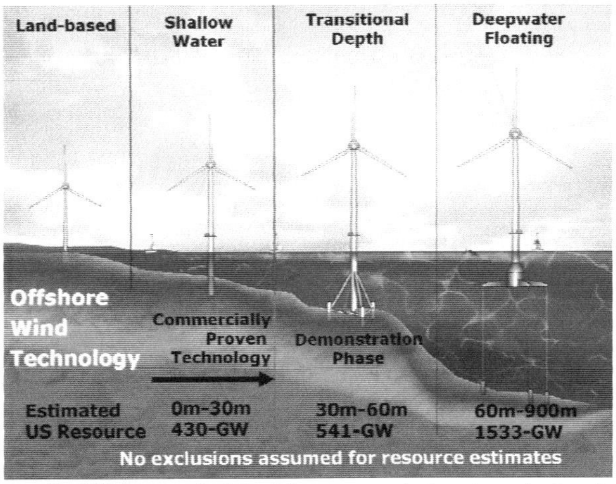

Figure 5: Offshore wind turbine support structure technology

1992). Table 3 is a general representation of the GOM in terms of water depths.

Typical support structures for wind turbines and approximate applicable depths are shown in Figure 5. Even though is not shown in Figure 5, there are different types of support structures in each of the shallow, transition and deep water categories. After evaluation for hurricane conditions, current support structure technology might be used in the areas of the GOM up to water depths of 30 meters.

5.0 WIND TURBINE HURRICANE RISK IN THE GOM

As indicated in Section 3, current standard wind turbines may not be able to survive Category 3 or stronger hurricanes. Either special wind turbine designs or careful risk assessments should be performed prior to installation and operation of a wind farm in the Gulf of Mexico. Category 3 hurricanes are within the 50-year design requirement of IEC 61400 (IEC 61400-1, 2005 and IEC 61400, 2009).

In many cases, design of offshore wind turbine structures is not governed by rare extreme events such as hurricanes but rather by operational loads. Furthermore, the offshore design is not only governed by wind but by wave, surge and current loads. A joint probabilistic approach for these events is necessary as rare wind/wave loads are not perfectly correlated in time. Typically, the currently available 110 years of historical hurricane records are used as a base for hindcast models. For the operational wind/wave loads simpler, yet robust hindcast models exist based in numerical weather simulations validated with measurements from sparse locations.

Hindcasting is a methodology used to characterize hydrology over a given bathymetry provided certain atmospheric conditions. Two of the most critical parameters for hindcasting are wind speed and wind direction (i.e. the wind field). The "basic" hindcast model used for hurricane events only takes into account the information provided from public records and fit a mathematical representation to create a wind speed and direction profile in a synthetic coordinate system. With this wind field used as input, the wave, current and surge information is extracted from the hindcast model for any given location in the artificial coordinate system.

Today, advanced methodologies can be used to more accurately predict the joint wind and wave characteristics that will affect offshore wind turbines in hurricane-prone regions.

5.1 Wind

Using the archived hurricane public records of events since 1900, probability distributions are created for key parameters that characterize the intensity, size and motion of the hurricane. The parameters include latitude and longitude, central pressure deficit, radius to maximum winds, translation speed, heading and landfall location. From the probability distributions, a Monte Carlo approach is used to sample from each distribution and create a mathematical representation of the wind field for each simulated set of variables. A stochastic track model combined with an ocean mixing model is used to characterize the lifecycle of a hurricane over the ocean. For a given location, several thousand simulated probable hurricanes will pass close or over the location of interest thus creating a much larger spatial sample that provide estimates of wind speed and direction in the time domain. Factors such as humidity, atmospheric temperature, and sea surface temperature are considered to model the strengthening and weakening of the storm as they evolve in a given track. This approach is the accepted methodology for definition of design wind speeds for onshore construction in the hurricane-prone areas. An example of the type of wind intensity created with such a methodology is shown in Figure 6.

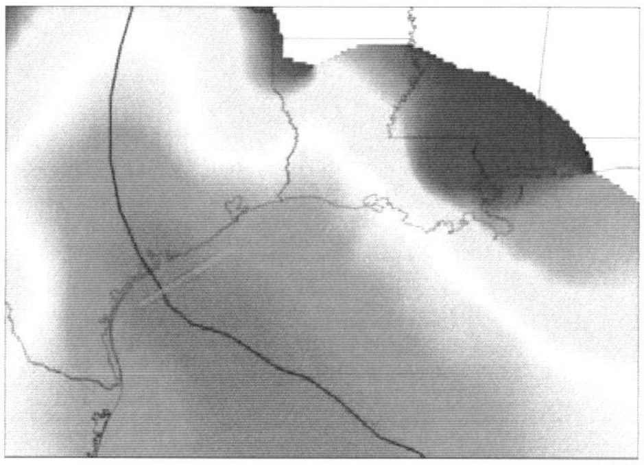

Figure 6: Hurricane sample modeled using ABS Consulting WorldCat enterprise

5.2 Waves, surge and currents

The wave, surge and currents are generally modeled with consideration for the bathymetry, ocean and atmospheric dynamics. A Finite Element approach is used to solve the incompressible governing equations (Shallow-Water Gravity equations). With the hurricane atmospheric stochastic set, WORLDCATenterprise (or other public or commercial software such as Mike 21) are used to model the hydrodynamic behavior of the ocean in shallow or deep waters in response to the time domain probabilistic hurricane models. A typical illustration, in this case surge height, is shown in Figure 7.

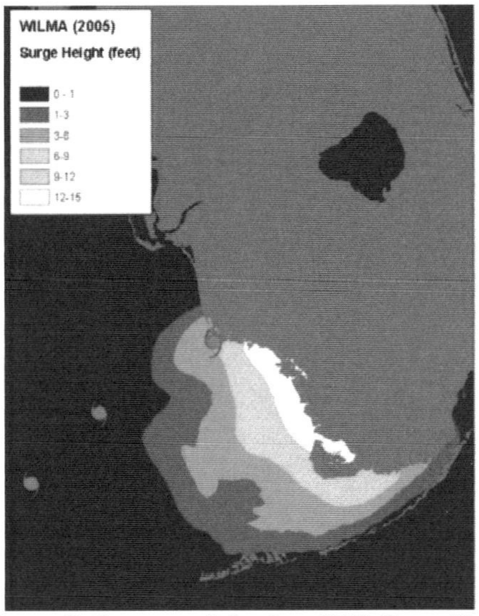

Figure 7: Surge model using ABS Consutling Worldcat enterprise

5.3 Coupled simulation

For a given set of hurricane wind speeds (and directions), respective wave heights, and spectral period, statistics are obtained for each time series. Conditional probabilities are examined: for example, maximum wind speeds are extracted along with the associated wave heights and directions at that time, as well as the maximum wave heights are extracted with the associated wind data at that time. With these conditional probabilities, a joint probability distribution is built where design requirements (i.e. 50- or 100-year design) can be established in a much more appropriate fashion than arbitrarily selecting different recurrence intervals for wind and wave as currently suggested with the "reduced values" in the International Electrotechnical Commission Standard (IEC 61400-3, 2009). Since the design of offshore wind turbines is

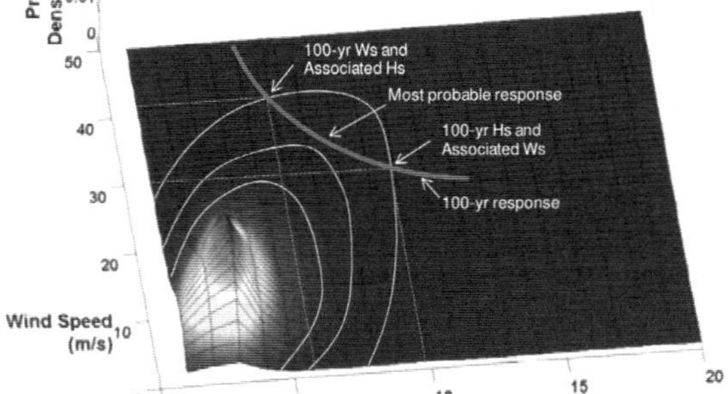

Figure 8: Joint wind/wave probability distribution

largely dependent upon the combined probabilities of wind and wave, defining a scientific methodology for the analysis is imperative to assure that wind turbine integrity is achieved, especially in hurricane-prone areas. Information from the coupled wind/wave statistical analysis will help to define design inputs for site-specific conditions as shown in Figure 8. At this point, two approaches are generally considered: (1) extrapolation of joint wind/wave (metocean) data to obtain design inputs for structural response analysis, or (2) the structure is analyzed for the full set of extracted hurricane atmospheric/ocean conditions from which extrapolation is carried out.

6.0 HURRICANE CONSIDERATIONS IN IEC 61400?

While IEC 61400 is an international standard, it is largely based on the experience from European wind farm development. Since there is not significant European exposure to hurricanes, there is considerable lack of specific guidance for these scenarios thus creating uncertainty regarding the behavior of wind turbines in US hurricane-prone areas. American certification bodies have the expertise to analyze and design for the effects of hurricanes in offshore environments – our history in the US is shaped by this fact!

7.0 ULTIMATE STRENGTH DESIGN

Ultimate strength design requires that a structure have the ability to withstand a pre-defined load without compromising structural integrity. When designing a structure to meet this requirement, several load cases involve limited duration simulations of stochastic sea states and turbulent inflow. From the limited duration simulations for the various combinations of wind and wave heights, statistics are obtained for each run such as to extrapolate the load effects intended for design, say 50- or 100-years.

For operational conditions, it is understood the need to carry out these multiple limited duration simulations to perform extrapolation of load effects (Fogle et al., 2008). However, for extreme event conditions such as hurricanes, analysis is typically a deterministic 50-year load effect for that set of wind and wave parameters obtained from extrapolation of met-ocean conditions. As the industry seeks to develop safe and yet optimized designs, extrapolation of combined hurricane wind and wave load effects is paramount.

8.0 THE ISSUE OF HURRICANE INTENSITY UNCERTAINTY

The MMS 618 report (MMS, 2009) presents a comparison of offshore wind turbines designed with the inherent reliabilities of two distinct standards, IEC 61400-3 (IEC, 2009) and API RP 2A (API, 2000). The conclusions suggest that the reliabilities with both standards are comparable despite the fact that IEC 61400-3 requires a design for 50-year event and API RP2A requires a 100-year design. Furthermore, the report suggests that design of wind turbines at higher latitudes, near the U.S. northeast coast, will use lower variations in the estimation of hurricane intensities as compared to those in the Gulf Coast. It concludes that lower coefficients of variation (variability) of hurricane intensities results in lower reliability of the offshore wind turbine system. The MMS 618 report shows coefficients of variation of 12% for the North Sea, while much larger values for the U.S. coast, between 25-53%. Clearly, the uncertainties in the MMS report for the hurricane-prone regions will result in lower reliability indices when compared to other European locations as uncertainty in the metocean estimates are twice if not three times larger. We believe that the hurricane intensity uncertainty may be comparable to events in Europe. There is no doubt that the wind intensity will be remarkably different between the two regions but hurricane intensity variations estimated in the MMS study seem to be an artifact of the metocean characterization method used. The author's argument is based on the fact that previous advanced hurricane models (Vickery, 2009), such as the one described in this paper, show considerably smaller coefficients of variation (c.o.v.) of hurricane intensity in the Gulf Coast (in the range of 6-7%), and coefficients of variation in the order of 12-16% in the U.S. northeastern tip. Lower uncertainty is found in the Gulf of Mexico given the more robust dataset of hurricanes available for the area. It is this author's opinion that the reliability of offshore wind power systems might be comparable to those in Europe if the systems is properly designed to the new site conditions (and accounting for wind directionality in hurricane events) in the U.S. coast.

9.0 CONCLUSIONS

Offshore wind will gain acceptance in the US through political support and by addressing perceived problems as challenges. Hurricane risk tools used for decades for the insurance industry can be coupled with publicly available wind turbine modeling software such as FAST [Jonkman, 2005] to provide a realistic reliability-based approach to the design of offshore wind farms in hurricane-prone areas.

Designs of offshore wind turbine systems in the U.S. that achieve comparable structural reliability of those in Europe are attainable. Proper metocean characterization in the hurricane-prone U.S. coasts is necessary to encourage development of an industry that is in its infancy and with lots of potential.

REFERENCES

[1] API RP 2A, Recommended Practice for Planning, Designing and Constructing Fixed Offshore Platforms-Working Stress Design, API RP-2A-WSD, 21st Edition, 2000.
[2] ASCE 7-05, Minimum Design Loads for Buildings and Other Structures, American Society of Civil Engineers, Virginia, 2006. p. 425.
[3] Cherry, N. J., D. L. Elliot and C. I. Aspliden, Worldwide Wind Resource Assessment. Proceedings of the Fifth Biennial Wind Energy Conference and Workshop, Vol. 2, National Technical Information Service, Springfield, Virginia, 1981.
[4] Elliot, D., Caribbean and Central America Wind Energy Assessment, Wind Power Conference, San Francisco CA, USA; Oct, 1987.
[5] Elliot, D., M. Schwartz, Wind Resource Mapping for United States Offshore Areas, AWEA WindPower 2006 Conference, Pittsburgh, Pennsylvania, June 4-7, 2006.
[6] Gore, R.H. 1992. The Gulf of Mexico. Pineapple Press, Inc. Sarasota Florida. 384 p.
[7] International Electrotechnical Commission (IEC), IEC 61400-1 Ed.3, Wind Turbines – Part 1: Design requirements, August 2005.
[8] International Electrotechnical Commission (IEC), IEC 61400-3 Ed.1, Wind Turbines – Part 3: Design requirements of offshore wind turbines, February 2009.
[9] Jonkman, J.M., M.L. Buhl, FAST User's Guide. NREL/EL- 500-38230. National Renewable Energy Laboratory; Golden, CO, 2005.
[10] Musial, W. and S. Butterfield, Energy from Offshore Wind, Offshore Technology Conference, Houston, TX, U.S.A. May 2006.
[11] MMS 618, Minerals Management Service, Comparative Study of Offshore Wind Turbine Generators (OWTG) Standards, Project No. 618, Prepared for JIP Sponsorship by MMI Engineering, July 1, 2009.
[12] Schellstede, H. J., Wind Power: Wind Farms of the

Northern Gulf of Mexico, Offshore Technology Conference, Houston, TX, U.S.A. 30 April-2 May 2007.

[13] Stewart S.W., US Offshore Extreme Wind Analysis based on Hurricane Return Probabilities, Poster presented at the AWEA WindPower Expo, Chicago, IL; May 2009.

[14] Vickery, P.J., et al., U.S. Hurricane Wind Speed Risk and Uncertainty. Journal of Structural Engineering (2009):135; 301-320.

[15] US Offshore Collaborative, US Offshore Wind Energy: A Path Forward, October 2009. www.usowc.org/pdfs/PathForwardfinal.pdf, Accessed January 5, 2009.

(This page intentionally left blank.)

WIND TECHNOLOGY MOVES INTO DEEPER WATER

Trond Landbø
Dr.techn.Olav Olsen a.s/Vici Ventus Technology, Lysaker, Norway
Nina Merethe Wiley, Arnstein Gorejord
Dr.techn.Olav Olsen Inc, Houston, TX, USA
Dagfinn Sveen
Dr.techn.Olav Olsen a.s/Vici Ventus Technology, Lysaker, Norway

ABSTRACT

The offshore wind industry has come a long way since the first wind turbines were set up off the coast of Denmark in the 1990s, but work still has to be done to make offshore wind energy more cost efficient as the challenges grow in terms of deeper water, increased turbine size, environmental loads, installation complexity, fabrication efficiency and operation stability.

The United Kingdom, Denmark and Germany are the leading nations in the development of offshore wind energy and are in the process of moving into deeper water. New innovations on turbine foundations are essential to keep the costs of deepwater wind farms from spiraling.

Dr.techn.Olav Olsen is a leading consultancy within structural and marine engineering with world-leading competence and experience in concept development and design of offshore concrete and steel structures.

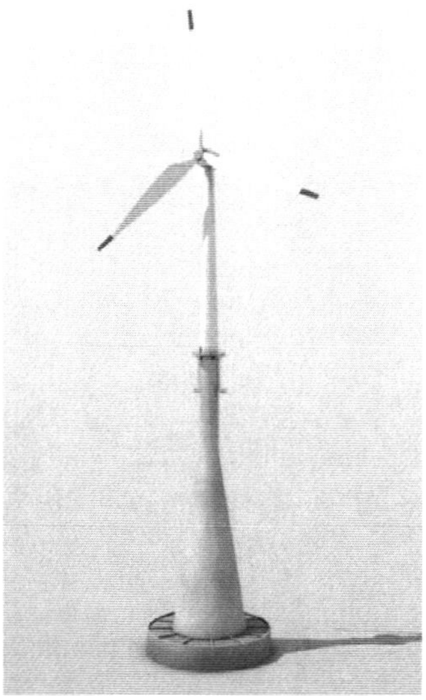

Olav Olsen has teamed up with AF Gruppen , a leading Norwegian contractor, and Lyse, a Norwegian Energy company to establish a new company called Vici Ventus Technology. Vici Ventus will focus on offering foundation solutions for the offshore wind industry and will draw on Olav Olsen's expertise in construction and engineering, the AF Group's strengths within contracting and project management, and Lyse's experience in developing and implementing major energy projects. Vici Ventus will make use of competencies that have evolved over several decades in the North Sea to develop, build and install cost-effective foundations for wind turbines.

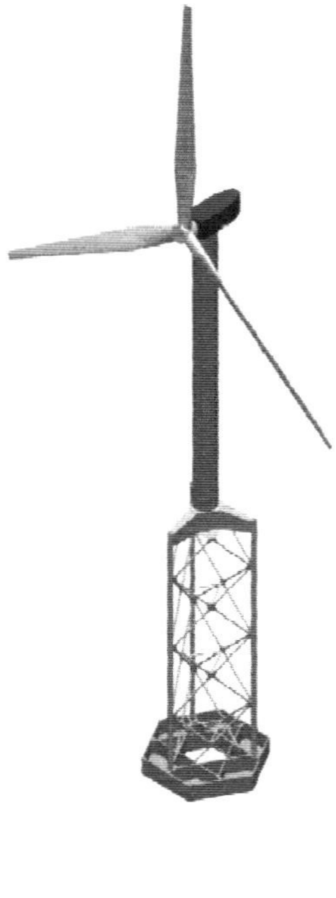

The paper presents Vici Ventus' patent-pending designs that are competitive for a variety of different water depths and seabed conditions, construction sites and installation methods:

Gravity based concrete substructure (CGS) based on well-proven technology from harsh environments around the world. The concrete substructure can be assembled inshore with the tower and turbine. In this way the use of expensive offshore cranes for installation is avoided. On arrival at the final destination, the base is installed through a simple ballasting operation. The concrete foundation provides a stiff and strong base for the wind turbine preventing unfavorable dynamic effects and fatigue problems. It can be designed for 100 years life or more, and require almost no maintenance. The solution is robust and durable, and significantly reduces risks and costs associated with transport, seabed preparation and installation.

The Space Frame Tower (SFT) is a unique wind turbine foundation concept combining excellent strength and fatigue properties with flexible and cost-effective fabrication. It is suitable for water depths up to 60 m. The triangular design with parallel legs and constant leg diameter enables efficient fabrication of components with equal lengths and angles – only member thickness is varied dependent on water depth and turbine type. Identical geometrical interfaces between all components make assembly simple and efficient. Fatigue is a critical issue for all offshore steel structures and for wind turbine foundations in particular. The standardized nodes – which can be mass-produced by hot-stamping and automatic welding – have excellent properties with regards to fatigue.

INTRODUCTION

New innovations on turbine foundations are essential to keep the costs of deepwater wind farms from spiraling. This paper presents cost efficient solutions for substructures in water depths up to 100 m. The concepts have been developed by Dr.techn.Olav Olsen. The concepts are Concrete Gravity Based Structures (CGS) and Space Frame Tower structures (SFT).

Dr.techn.Olav Olsen has together with AF gruppen, a leading Norwegian contractor, and Lyse, a Norwegian Energy company, established the company: Vici Ventus Technology.

Dr.techn.Olav Olsen (OO) is an independent structural and marine engineering company working with projects onshore and offshore. It was founded in 1962 at Lysaker, Norway and in 2006 it opened an office in Houston, USA. The company is engaged in consulting, design and engineering for the Norwegian and International Oil & Gas Industry, Civil construction and Transportation.

OO has experience from offshore projects, bridges, quays, commercial buildings, residential buildings, schools, sports arenas and other public buildings.

OO was central in development of the Condeep platform in the early 1970s and the company has designed more than 60% of offshore concrete platforms worldwide. OO introduced skirt piling in early 1980s as solution for soft soil and Gullfaks C, developed and designed Troll A GBS for 305 m water-depth. Developed and designed Heidrun TLP concrete hull and foundations. In late 90s OO developed Heavy Lift vessels for installation/removal of large offshore structures, including wind turbines. Latest offshore concrete projects are detailed design of Sakhalin 2 platforms (LUN-A and PA-B) and the ALT LNG terminal, both projects for Aker Solutions. Detail design of concrete GBS for Sakhalin 1 is an ongoing project.

OO has a long record in design of various offshore steel structures, usually as a subcontractor to large international contractors. Typical projects include design of topside structures, jacket structures, jack-up structures, semisubmersible hull structures, FPSO hull structures, flare booms, suction anchors, lifting beams, riser support structures etc. OO has also taken full engineering responsibility in projects involving mechanical outfitting, piping, E&I and marine engineering.

The company has significant experience in project administration, such as project management, design management and construction management. The services are approved by the National Office of Building Technology and Administration within our service fields.

The company actively participates in research and development within its field, and has received several acknowledgements from clients and learned societies for exceptional professional merit. In 1976, 1984 and 1985, the firm was awarded the Norwegian Betongtavlen, in 1990 and 1998 FIP's prize for outstanding structures. In 2000 the readers of Teknisk Ukeblad elected Troll A, a large offshore platform developed by Dr.techn.Olav Olsen a.s for Norwegian Contractors, the engineering achievement of the century. In 2007 the company's founder, Olav Olsen, was awarded a Technology Pioneer by the Offshore Energy Center Hall of Fame in the category: Production Technology and Concrete Gravity Structures for his outstanding contributions to the offshore industry.

Specific competence is aimed at structural consultancy on areas of extraordinary technical challenges – specifically related to alternative design and construction methods, structural systems and foundations. Typical assignments include industrial multidisciplinary projects, for Norwegian and international clients, where interface co-ordination and management may be part of the assignment

Our markets are all over the world such as the North Sea, Far East, Australia, Western Africa, Gulf of Mexico, Eastern Canada and the Barents Sea. We have designed platforms for the most severe seismic, wave and ice exposure.

Renewable energy is defined as a focus area, with special focus on offshore wind. Over the last few years OO has performed a number of studies for different clients. These are studies related to design and fabrication of fixed or floating offshore wind turbine foundations (concrete and steel) as well as aspects related to offshore marine operations, heavy lifting, O&M etc.

Vici Ventus Technology AS was founded in September 2009 by the 3 equal partners AF Gruppen ASA (www.afgruppe.com), Lyse Produksjon AS (www.lyse.no) and Dr.techn. Olav Olsen a.s. The 3 partners have joined forces to bring their complementary knowhow to the offshore wind power industry. The company will draw on Olav Olsen's expertise in construction and engineering, the AF Group's strengths within contracting and project management, and Lyse's experience in developing and implementing major energy projects.

Vici Ventus has an overall objective to deliver cost effective foundation structures to the offshore wind industry. The range of products and services includes conceptual design and services related to early phase studies, detail engineering, fabrication and installation of foundations at offshore site. Vici Ventus has access to one of the most modern fabrication sites in Europe, the AF Miljøbase Vats owned by AF Gruppen ASA. This is the site where large platform structures, such as Gullfaks C, Draugen and Troll A, were built and assembled.

Vici Ventus has designed two unique, patent-pending substructure solutions for offshore wind turbines; One Concrete Gravity Structure (CGS) and one steel Space Frame Tower (SFT) structure. Both concepts are very cost effective and each has unique characteristics making them to preferred solutions in different water depths and areas of the world.

CONCRETE GRAVITY BASED STRUCTURES

Introduction

The Concrete Gravity Based Structure (CGS) has been developed by Dr.techn.Olav Olsen a.s on behalf of the technology owner Vici Ventus Technology AS. Patent applications have been filed in 2008 and the concept is "patent pending".

Previous studies have proven that concrete gravity base foundations for offshore wind turbines have favorable structural characteristics and behavior, and are very competitive in medium to deep waters.

The foundation concepts may support large offshore wind turbines due to high strength and robustness; also the 10-15 MW turbines for the future. The concept is designed for rough conditions such as the North Sea.

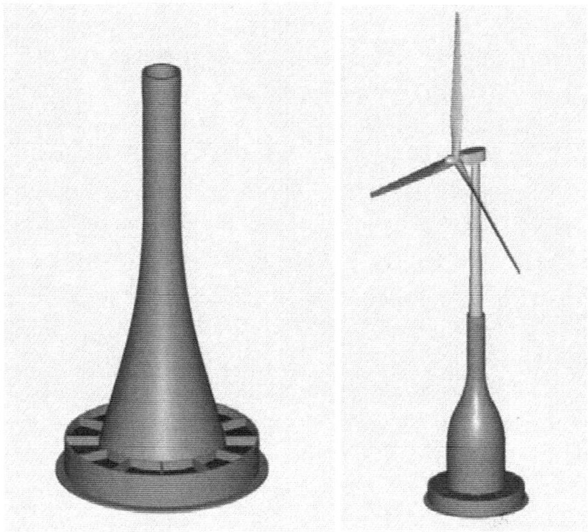

Figure 1. Concrete Gravity Based alternatives for 50 m water depth. Offshore installation of wind turbine generator (left) and complete inshore assembly (right)

Concept Description

The concrete gravity based structure concepts presented here has been developed based on various parameters. The main variables have been:

- 5 MW wind turbines
- Water depth: 30-100 meters water depths have been investigated. North Sea conditions.
- Inshore and offshore installation of wind turbines (WTG e.g. nacelle, tower and blades).
- Utilization of different types of Normal Density (ND) and Light Weight Aggregate (LWA) concrete.
- Use of different solid ballast materials
- Different combinations of inshore and offshore filling of solid ballast.

The main objective has been to develop cost effective foundations, and one aspect of that has been to determine the range of water depth when inshore installation of the WTG is cost effective. The conclusion is; water-depths of 45 m and deeper. The major advantages of inshore installation are reduced risk and cost, because lifting operations are moved to sheltered areas less affected by wind and waves.

In 30-40 m water-depths however, offshore installation is the preferred choice. One reason is that in water-depths up to 40 meters the supply of jack-up crane vessels seems reasonable. Going into deeper waters than 40 meters; feasible installation vessels become scarce and offshore installation of wind turbines may be very expensive, and may also include high commercial risk. Temporary buoyancy elements for concrete CGSs in 30-40 meters water depths have been briefly evaluated as a mean to achieve inshore installation of WTGs. This method however, appears to be more expensive than sizing up the foundation itself.

In the design, a normal high strength type concrete will be used, with typical cylinder strength 50-65 MPa. Such concrete can be produced with different densities, and the density of Light Weight Aggregate (LWA) and Normal Density (ND) concrete including steel reinforcement typically ranges between 1,6 and 2,7 t/m^3. It may be optimal to use a combination, typically with LWA concrete in the shaft and ND concrete in the base.

For water-depths 30-60 m there will typically be a requirement for solid ballast filling inshore, but also for adding solid ballast offshore. The requirement for offshore solid ballast will be reduced with increased water depth.

Typical design life for such a CGS is 50-100 years, meaning it can be reused at the site with a new turbine on top, or it can be relocated to a new site with similar water depth and soil conditions. The CGS can be removed in a simple way by reverse installation and there will not be any structural elements remaining on or below the seabed.

There are several possible solutions for connection between the CGS and the steel turbine tower. The figure below presents a solution that allows for wide fabrication tolerances and a robust and fatigue resistant connection which fits well into the fabrication schedule. The bolt connection is kept inside the tower, protected from sea spray etc.

Concrete structures are not sensitive to fatigue, and use of a CGS will almost eliminate the issue of fatigue for the foundation structure itself. The only part of the foundation structure which will be sensitive to fatigue is the top steel flange which is easy to make robust also with future extension of operation or reuse in mind.

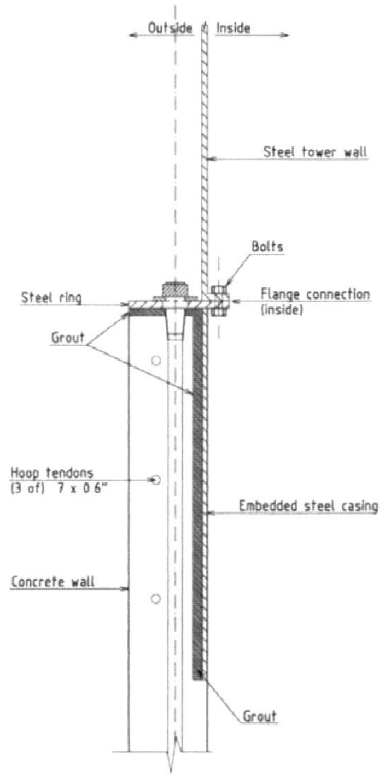

Figure 2. Connection between CGS and steel tower

Soil Conditions

The CGS concept is based on proven technology for large variations of soil conditions. The majority of such structures is installed on dense sand or firm clay, and has short ribs/skirts only for scour protection and load distribution.

For soft soil, longer skirts made of steel or concrete is introduced. -In the mid 80's Dr.techn.Olav Olsen introduced skirt piling for the Gullfaks C project, and since then the skirt piling method has been used for a number of platforms: Troll A is the extreme case where the skirts penetrate 36 m into the soft soil. The skirt piling concept has been further developed for suction anchors and bucket foundations. Use of skirts will typically reduce or eliminate the need for solid ballast.

Normally no preparation of the seabed is required, and a sloping seabed can be accounted for in the design. The foundation structure is robust and capable of resisting uneven soil pressures due to small piles of sand etc. When selecting site, effort should be made to avoid large sand waves or gravel piles.

Fabrication

The construction of the concrete CGSs for wind turbines will basically follow the same well known and proven techniques, procedures and methods as developed through the fabrication of concrete offshore platforms; especially the Condeep platforms in the 1970s, 80s and 90s in Norway.

The bottom section of the substructures will be constructed either in a dry dock or on a semisubmersible (SS) barge moored along a key side. Ordinary formwork will typically be used, and for mass production, effort will be made to prefabricate standard formwork elements for multiple re-use. The fabrication cost of a mass produced units will be 50% less than for an initial single unit.

Alternatively, the units can be built on a quay, and the base can be skidded over to a SS-barge for launching. The maximum floating draft of the bottom section is determined by the available draft in the dry dock, or the submersability of the SS barge. To make it cost effective, effort has been made to limit the floating draft to a minimum, typically 5-6 meters. Fabrication of one bottom section takes typically 3-4 months. For mass production on an established fabrication site, 3 months is a viable estimate.

When afloat, the bottom sections will be towed to a slip-forming site with a suitable water-depth. At the slip-forming site, the concrete shaft will be completed while the substructure is afloat and moored to a utility barge. Post tensioning and mechanical outfitting will be performed at the slip-forming site. Completion of the concrete work at the slip-forming site takes typically 4-6 weeks. In order to complete the concrete structure, a minimum water depth somewhat less (about 10 m) than the water depth at the offshore site is required.

Installation of solid ballast at the inshore site will take place by direct filling from the ballast transport vessel to the CGS through a hopper and a fall pipe arrangement while the CGS is moored at the slip-forming site. Possible types of solid ballast are:

- Sand (typical wet density 2.0 t/m^3)
- Olivine with wet density 2,425 t/m^3 (may achieve 2,65 t/m3 when ballasted dry)
- Iron ore (typical with wet density 3,5 t/m^3)

In cases when the wind turbine and its steel tower are installed inshore, these components will be assembled and lifted on top of the CGS substructure while the CGS is moored at the slip-forming site. For this operation a water-depth similar to the offshore site is required. If larger water-depth is available, the lifting height for the tower and turbine can be minimized.

As offshore installation will normally not take place during the winter months, sufficient buffer storage area for completed substructures during this period needs to be provided for.

The figure below shows a typical constructions schedule for one CGS. This is a typical schedule for a pilot project with the aim to prove the concept, the fabrication methods and the cost potential. The fabrication of the base structure is scheduled to take 4 months, marine operations and slip-forming of the shaft 2 months, completion and installation 2 months. With a general rigging period such a pilot project will take about 10 months.

A simplified fabrication schedule for mass production is shown below. Activities for the first year of production will be repeated the next year, and are for simplicity not shown. Total duration of the EPCI contract period for the 96 foundations is assumed to be 36 months, including preparations and demobilization.

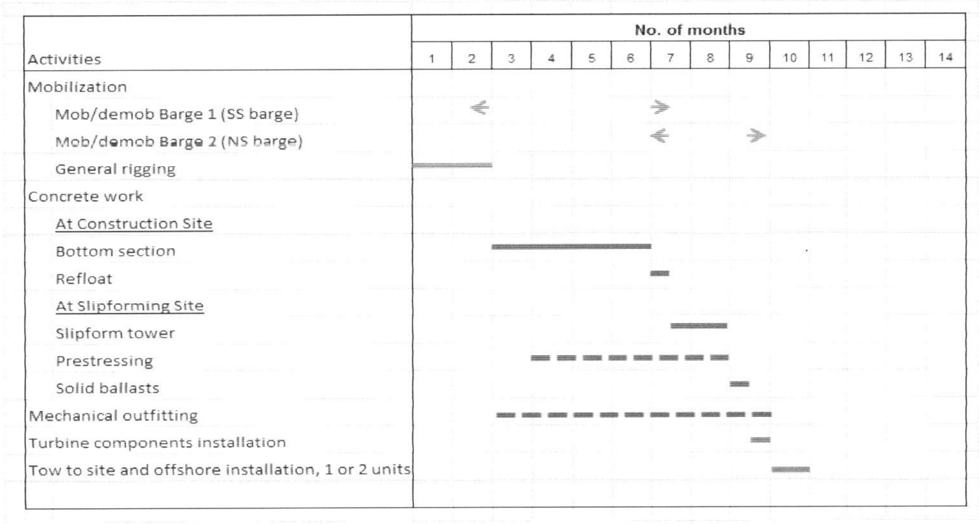

Figure 3. Single turbine fabrication schedule

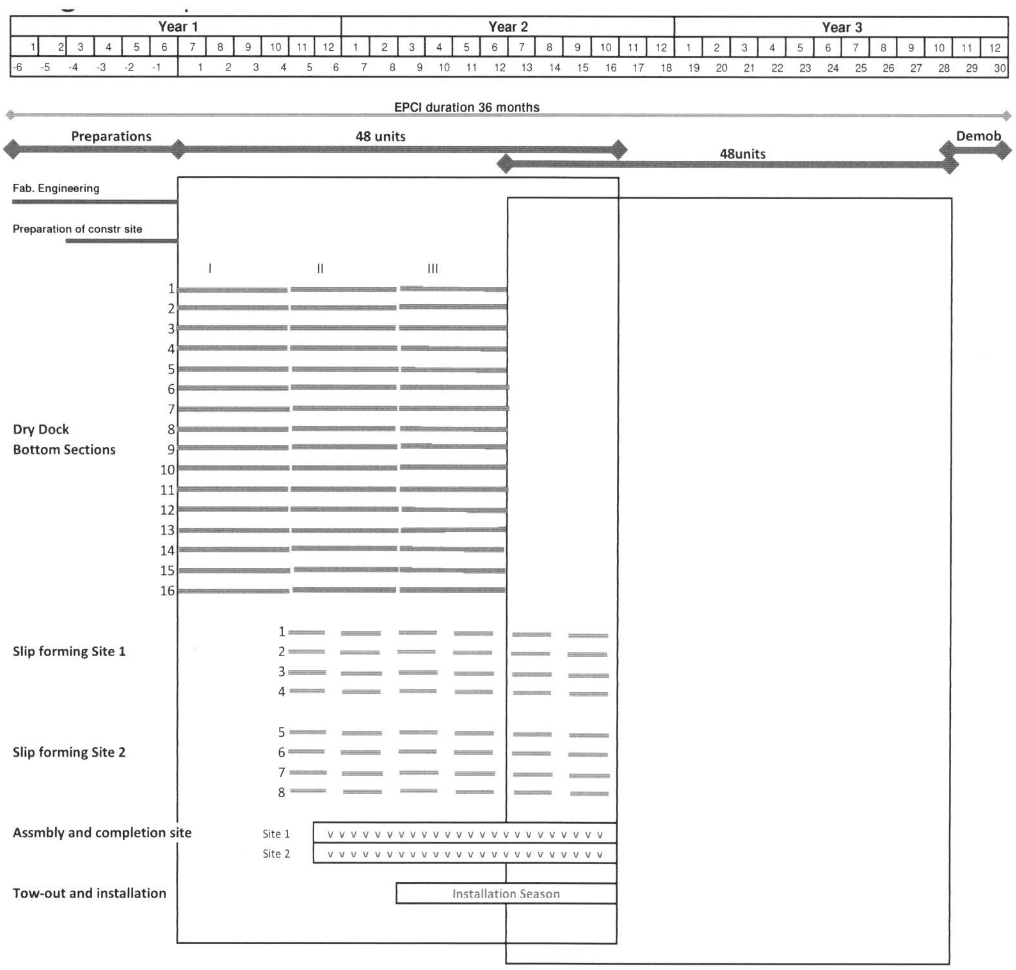

Figure 4. Large scale production of 96 units during 2 seasons

Installation

The CGS will be towed from the fabrication/assembly site, fully assembled with WTG or as a separate complete foundation unit.

After completion of the fabrication the unit will be hooked up with a towing spread. The towing spread for towing one CGS to the field will typically consist of one large towing tug (150-200t bollard pull) and two smaller tugs (50t bollard pull).

Arriving at site, the installation procedure is to position the towing spread with the CGS over its final location and ballast with water until the CGS is resting on the sea floor. The towing spread is then disconnected, and will return to the construction site for another CGS. If required, more solid ballast is filled to make the CGS stable for all conditions over its life time. Prior to such filling, the foundation will always be able to resist a summer storm condition, and hence this operation can be done as a combined filling operation for several foundations at the end of the summer season.

The CGS base is provided with a steel skirt to prevent scouring. Preparation of the sea bed is normally not needed, neither prior to nor after the installation. Effort will be made in the design to avoid under-base grouting in order to save cost and time, and in order to simplify removal and a possible relocation of the foundation structure.

Removal of the CGS is done by reversing the installation operation. If offshore ballast was installed after installation this must be removed prior to de-ballasting and lift-off. A controlled lift-off is achieved by pumping seawater into the skirt compartments.

For a fully assembled units (typically in water depths 45 m and deeper) the wind turbine is already installed on top of the CGS, and the wind turbine is ready to be hooked up to the electrical grid.

Otherwise, if CGSs are installed separately (typically in water depths less than 45 m) the wind turbine will be post installed offshore by a special installation vessel for turbines.

Summary

Key features for the CGS foundations are:

- No use of high-cost installation vessels for foundation installation. Lowering to the seabed by water filling only.
- Simple removal by reverse installation method.
- With inshore installation of the wind turbines critical lifting operations offshore with associated high costs and high risk exposure will be eliminated.
- Preparation of the sea bottom prior to or after installation is normally not needed. Inclined seabed and local level-variations can be accounted for in design.
- Durable concrete foundation, design life time 100 years or more. The concrete structure is robust and not sensitive to fatigue. Possibility for reuse; on same site with new turbine or alternatively on a new site with comparable water depth.
- Excellent dynamic behavior. The concrete structure is normally very stiff in all degrees of freedom and it is simple to tune the natural periods of the turbine by adjusting the stiffness of the foundation or the steel tower.
- The bottom section of the CGS foundation can be constructed in a dry dock or on a semisubmersible barge. The main tower is slip-formed while floating. A sheltered and deep water site is required for the slip-forming and the full assembly.

SPACE FRAME TOWER

Introduction

The Space Frame Tower (SFT) is a light steel structure with very high bending stiffness and superior fatigue properties that makes it an excellent foundation for big wind turbines in water depth from shallow to deep (60m). The structure is very simple and is based on a few types of standardized elements that build up the complete structure. Fabrication is based on subcontracting the structural elements to various fabricators that can specialize on single element types. This will ensure unlimited fabrication capacity, high quality, high efficiency and low cost. It also makes it possible to engage smaller fabricators. The SFT structure will be assembled at sites close to the offshore wind park to minimize transport time for installation. Assembly will be fast and cost effective as the assembly work is simple and based to a high degree of automated welding. Installation is based on barge transport and lifting operations. It is, however, envisaged that more efficient specialized installation vessels will be developed for future mass installation of wind parks. Different foundation solutions can be used depending on soil conditions, economical considerations, availability of installation equipment etc. Solutions as concrete gravity base, piles, suction piles have been considered. The basic idea with the SFT concept is to provide a standardized design that can be used at almost any wind farm site, and by its cost effective design and fabrication will be the preferred solution.

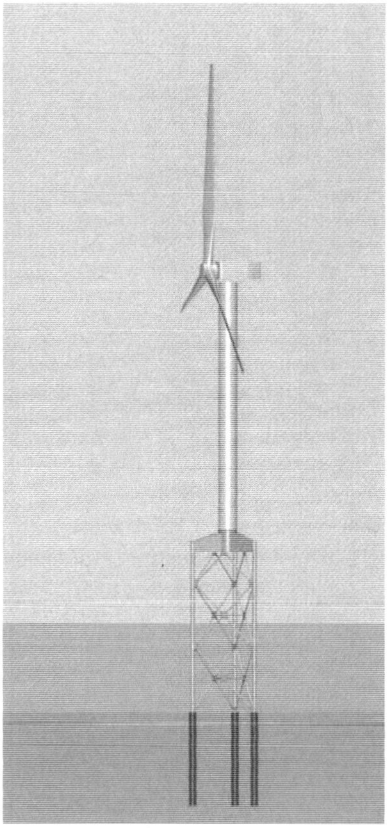

Figure 5. Space Frame Tower

Concept Description

The space frame is a triangular shaped structure with constant/parallel leg spacing. The braces are arranged in an X-pattern and the angle will be dependent on the leg distance and the tower height given the requirement that the angle shall be constant for all the braces. The angle may then vary between 40 to 49 degrees.

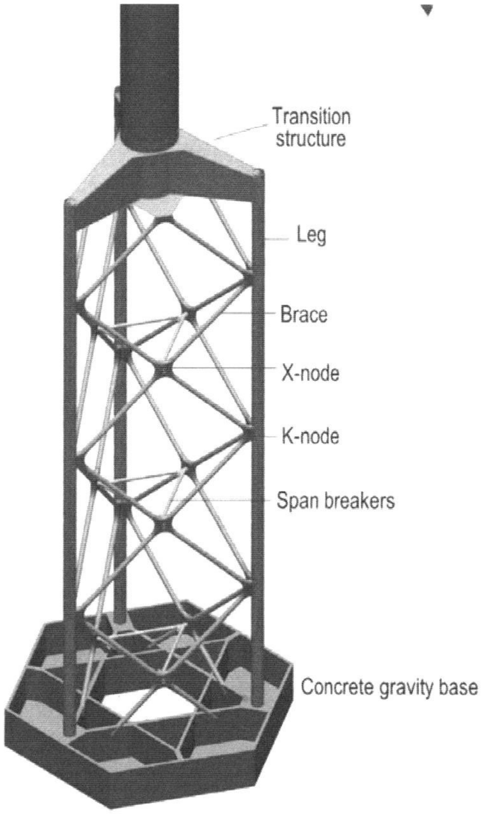

Figure 6. Space Frame Tower (SFT) Concept

The X- bracing is structurally very efficient and gives a structural redundancy as it is constituted by two independent load patterns.

To secure the stability, both in intact and damaged condition, an internal "span breaking" brace system is arranged. This is important both for the stability and to reduce the local wave load bending effect on the braces.

The legs have constant outer diameter. The required local stiffness and strength properties are tuned by the local wall thickness. The main braces have also equal and constant diameter. This makes it possible to use only one type of nodes that can be efficiently mass produced by a specialized fabricator.

For a given turbine size and make it is assumed that the basic design can be standardized, independent of water depth. This means that the same dimensions are used for the nodes, brace diameters, leg diameters and leg spacing. Only the height of the tower will be tuned to fit the water depth and the local wave heights.

The transition structure connects the tower with the space frame structure. It is a built plate structure with high overall stiffness and good design details to avoid fatigue.

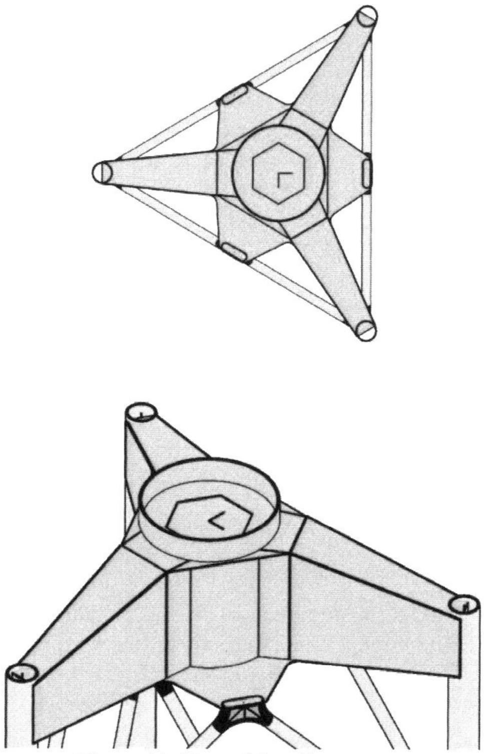

Figure 7. Transition Structure

The transition structure, like the turbine tower, is primarily dimensioned by the wind loads and the dynamic response in the turbine. Accordingly, for a specific turbine size and make, it is assumed that the transition structure can be standardized by keeping one leg distance for all water depths.

The access platform will be arranged at the lower level of the transition structure with a door through the central cylinder wall.

The nodes are of a special design with superior fatigue capacity, and they are exceptionally suitable for cheap and efficient mass production. The proposed design with parallel legs and X braces will have only one type of nodes both for the K-nodes on the legs, and the X-node for the braces. There will be only one brace diameter such that all nodes will have equal dimensions. Two K-nodes are made by cutting an X-node in two parts.

The nodes have been used in Jack-up platform legs where several hundred equal nodes are required. The nodes have been mass produced by hot forming precut steel plates which have been edge prepared and joined by a central automatic weld.

For the present design the brace diameter is 500 mm and the node thickness is 30 mm. Weight of one K-node is approximately 1100 kg and the X-node 2200 kg.

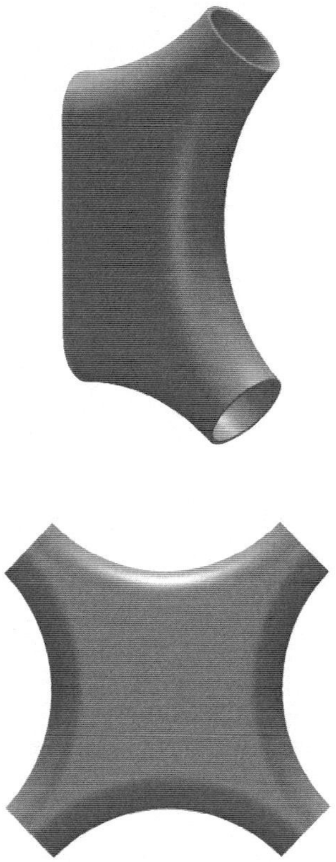

Figure 8. K-node and X-node

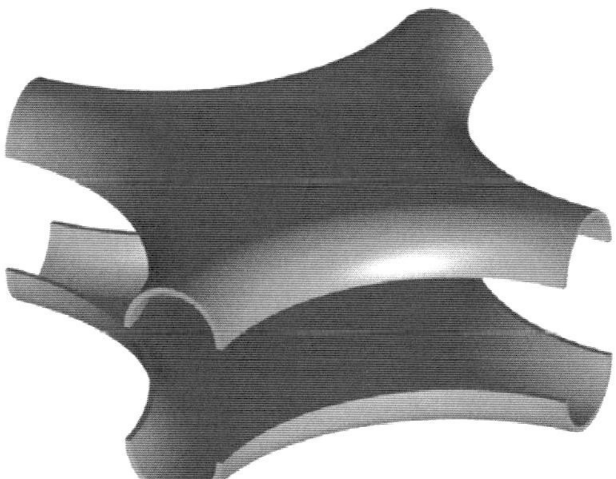

Figure 9. X-node

By using these nodes the assembly of the structure will be very efficient and simple as the welds will be braced to stub connections, and a geometrical simple and non critical weld between the K-node and the leg. No increased thickness in the chord plate is required.

At each X-node level there is an inner triangular brace arrangement, so called "span breakers."

The span breakers stabilize the X-braces and reduce wave induced bending stresses. The span breakers are also important in a possible damage condition where a brace has been broken or bent from i.e. a collision with a service vessel.

The span breakers will be connected to the X-brace by a lateral stub that goes through the node.

Seabed support

The seabed support solutions can vary dependant on soil conditions and other conditions related to fabrication and installation aspects. For the SFT the following foundation solutions have been considered possible:

- Concrete gravity base
- A piled support frame in concrete or steel
- Steel piles through side connected pile sleeves
- Steel piles through the legs
- Pre installed steel piles by use of a guide frame
- Bucket suction supports directly to the legs
- Bucket suction support with a support frame

The foundation solutions can be grouped into three main solutions: Gravity based, hammer piled and suction (bucket) piled.

The concrete gravity base is optimal from the view point of minimizing size and weight. This is because the space frame structure attracts low wave forces and accordingly minimizes the over turning moment on the foundation compared with big diameter columns.

The reason to use a hammer piled solution can be weak soil conditions and an uneven sea bottom.

Suction or bucket anchors are commonly used in soft clay and sand conditions. Use of suction anchors does not require utilization of costly piling vessels.

Fabrication

The fabrication of the SFT is divided in two main sequences:

- Fabrication of the main parts
- Final assembly of the main parts

Fabrication of the main parts can be spread to various sub contractors and will include

- Tubular members for legs and braces
- Nodes
- Transition structure

The assembly of the main parts will consist of rather simple welding work that can be performed partly with robots. The assembly site should be close to the wind farm site to minimize transport time and cost.

Bottom support solutions may include concrete structures, preferably fabricated on the SFT assembly site. These concrete structures are rather simple and shall be built on flat land and transported to the quay front for either transport to the wind farm site for separate installation or mated with the SFT, if it shall be installed as a complete unit.

The concrete structures can have weights from 600 to 1200 tons, and for mass production it is envisaged that the fabrication is arranged along a common skid way that leads to the quay front or the structures can be transported on trailers that are commonly used in offshore fabrication.

A typical schedule for the construction of one SFT foundation with concrete base structure is shown in the figure below. For large scale production, the duration of the manufacturing activities are expected to be substantially reduced.

Installation

Installation can be performed in many different ways dependant on various factors as foundation solution, available equipment, number of units, economical aspects etc. For a pilot installation available methods and equipment will be used, while for future wind parks special installation methods and vessels may be developed to increase efficiency and to reduce cost.

The SFT may, as described, have 3 different foundation solutions; gravity based, hammer piled and suction piled (buckets). All solutions can be arranged as pre-installed foundations, and the SFT is installed onto the foundation by lifting. The SFT can then either be mechanically fixed or grouted to the foundation. The SFT can also be installed by lifting with the foundation already mounted. In the case of a gravity foundation the lift weight will then be rather high (1500 to 2000 tons) and may require a rather expensive crane barge. However, this may require less marine operations compared with using preinstalled foundations. In a wind park scenario with a great number of installations there will probably be more beneficial to preinstall foundations as this work can be arranged in a very effective way with optimized equipment and planning. With pre-installed foundations the installation of the SFT can then be done rather quick and safe, especially if mechanical locking to the foundation is used.

Summary

Key features of the steel Space Frame Tower foundations are:

- For a given turbine size and make, the design is standardized for variable water depths
- K- and X-nodes are only of one design
- Simple fabrication method
- Several support solutions to choose from depending on soil conditions and fabrication & installations aspects
- The design facilitates use of various sub contractors for fabrication of main parts
- Pre-installed foundations may be advantageous for mass production

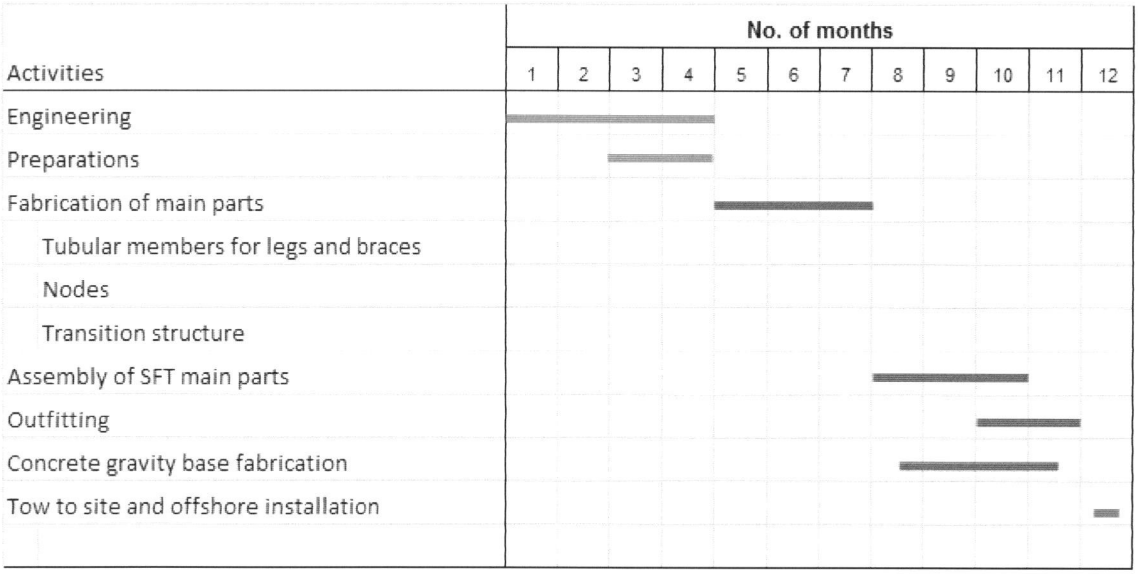

Figure 10 Schedule of SFT foundation with concrete base

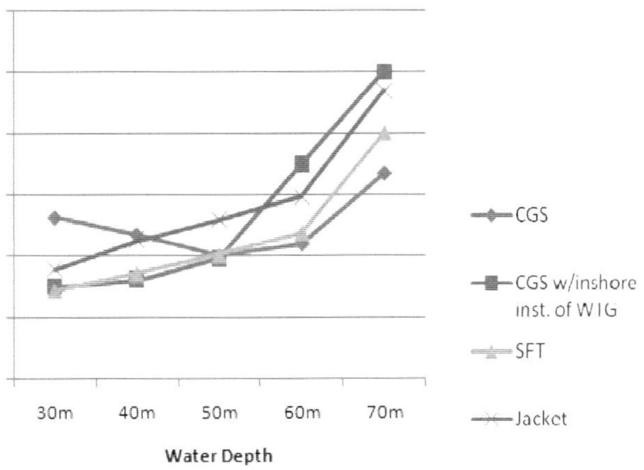

Figure 11. Trend graph of total cost including WTG installation

COST COMPARISONS

For benchmarking purposes, the cost of a traditional jacket substructure is compared with the presented concepts; the CGS and the steel SFT. The results show that both concepts can compete with the jacket design for all water depths. The CGS with capacity of pre-installing the wind turbine generator inshore is the most favorable solution for deeper waters.

CONCLUDING REMARKS

The wind turbine foundations presented herein presents cost-effective and competitive solutions for water depths up to 100m. The concepts are based on well proven technology, and are standardized to accommodate mass production. The concepts can be designed for all types of soil conditions, and normally no seabed preparation is necessary. The concepts are competitive to traditional jacket structures. When inshore installation of the wind turbine generator is feasible, the CGS concept is the most favorable with respect to cost and risk.

(This page intentionally left blank.)

Proceedings of the 16th Offshore Symposium, February 9, 2010, Houston, Texas
Texas Section of the Society of Naval Architects and Marine Engineers

TENSION LEG PLATFORM TURBINE: A UNIQUE INTEGRATION OF MATURE TECHNOLOGIES

William L. Moon III (AM) and Charles J. Nordstrom (AM)
The Glosten Associates, Seattle, Washington

Concept rendering of a tension-leg platform supporting a 5-MW offshore wind turbine in 55 m water depth (fish-eye view) – The Glosten Associates

ABSTRACT

The TLTP (Tension-Leg Turbine Platform) is a floating, permanently moored structure designed to support a utility-scale offshore wind turbine in water depths of 40 meters and greater. This paper discusses the design tools and unique design process used for this novel structure. Reasons for selecting a tension-leg platform (TLP) are discussed. Motivations to apply TLP technology to relatively shallow water depths are explained. The requirements for supporting a wind turbine, which are fundamentally different than the requirements for supporting a typical oil and gas payload, are discussed.

KEYWORDS

Wind turbine; floating; tension leg platform; chain mooring; design spiral.

INTRODUCTION

The Tension Leg Turbine Platform (TLTP) is the integration of many mature technologies including tension-leg platforms (TLP), offshore utility-scale wind turbines, permanent chain mooring systems and high-vertical-load anchors. Together, they form a unique solution to the intermediate and deep water[1] challenges inherent in the next generation of offshore wind farms. This paper provides a general overview of our TLTP design concept, a description of the environmental and design criteria, and a discussion of the reasons for choosing a three-armed tension leg platform.

The TLTP is fundamentally different from offshore oil and gas tension leg platforms in its payload, loading, target water depth, and tendon materials. Siting the TLTP in water depths of 40 to 100 meters requires the mitigation of design challenges not fully realized by either traditional ship or tension-leg platform design processes. Consequently, our unique TLTP design process, along with a discussion of the tools used for design and analysis, is presented as well.

DESIGN CONCEPT OVERVIEW

The TLTP design concept is based on requirements for a generic 5-MW offshore wind turbine to be sited in environmental conditions found in the United Kingdom Round 3 Development Area. As with most wind turbine foundations and offshore platforms, this design is highly site-specific. Offshore wind turbine characteristics, environmental criteria, and details of the supporting tension leg turbine platform are discussed below.

Wind Turbine Characteristics

The offshore 5-MW wind turbine used for the present design is a hybrid between the REPower 5-MW and Multibrid 5-MW turbines. It is similar to the 5-MW NREL Reference Wind Turbine described in Jonkman, *et al.* [2], but with a heavier nacelle/rotor and a shorter, and thus lighter, tower. The interface between wind turbine tower and hull is 18.0 meters above the mean sea level (MSL). The hub height is 95.5 m above MSL. As shown in Figure 1, a 5-MW TLTP positioned next to the Seattle Space Needle would be comparable in height.

Principal characteristics of the wind turbine are provided in Table 1.

For the coupled dynamic modeling discussed later in this paper, the 5-MW NREL Reference Wind Turbine, as described in Jonkman, *et al.* [2], is used for definition of controls and operating parameters.

Environmental Criteria

The offshore wind industry has, in general, chosen a 50-year return period to define the extreme environmental conditions. This differs significantly from the offshore oil and gas industry, which uses a 100-year return period. As onshore and offshore wind turbines are unmanned and typically have a shorter design life, the ultimate limit state return period can be lowered.

Figure 1: 5-MW TLTP and Seattle Space Needle

Table 1: 5-MW offshore wind turbine characteristics

Turbine Characteristics		
Tower Weight	300	tonnes
Nacelle and Rotor Weight	410	tonnes
Hub Height above Tower Interface	77.5	m
Rotor Diameter	126.0	m
Cut-Out Wind Speed	25	m/s

The specific location of interest in the UK Round 3 Development Area has a 50-year, 3-hour significant wave height of 8.5 meters, with an associated peak period of 12.5 seconds. Wave loads are modeled using a JONSWAP sea spectrum with a peak enhancement factor of 3.3. The 1-year statistics are listed in Table 2 for the site of interest.

For operational (wind turbine operating) load cases, significant wave height and peak periods are developed as a function of the wind speed taken at 10 m above the sea surface. For example, the cut out hub wind speed of 25 m/s has a significant wave height of 4.5 meters, and a peak period of 9.0 seconds. The correlation is based on ten years of hindcast data.

The 50-year, 3-second gust, at 100 meters above the sea surface, is 55 m/s. Using logarithmic scaling, this correlates to a 1-hour average wind speed of 33 m/s, at a height of 10 meters above the sea surface.

The turbulence intensity for the maximum wind condition is 12%, which gradually increases to 20.7% as the wind speed approaches zero. The maximum wind speed standard deviation is, therefore, 5.6 m/s.

The Kaimal spectrum, as recommended by DNV OS-J101[2], is used to develop the spectral density.

For wind spatial coherence, the Davenport spectrum is used as described in DNV RP-C205. This spatial variation is used to calculate the yaw moment by modeling each blade with varying instantaneous wind velocities.

Current speed is taken as 4% of wind speed at the sea surface, which is typical analytical practice for a Gulf of Mexico hurricane. A vertical profile of the current velocity is implemented.

For ultimate limit state analysis (50-year conditions), wind, waves, and current are assumed to be collinear. Operational states consider both aligned and misaligned wind, waves, and current.

Water depths for UK Round 3 sites range from 35 to 60 meters. The chosen water depth for this TLTP design is 55 meters. Water depths as shallow as 40 meters are possible with the TLTP, but a more benign wave environment and/or a smaller wind turbine is needed in comparison to the design presented here.

Table 2: Environmental summary

Environmental Summary		
H_s / T_p (50 year, 3 hr storm)	8.5	m / 12.5 sec
H_s / T_p (1 year, 3 hr storm)	6.7	m / 11.1 sec
U_{100m} (50 year, 3 sec. gust)	55.0	m/s
U_{100m} (50 year, 10 min. avg)	47.0	m/s
Current (surface)	1.3	m/s
Tidal Variation	4.0	m
Water Depth	55.0	m

TLTP Overview

The TLTP hull is comprised of several main components, which are shown in Figure 2 and described as follows:

1. Upper Hull – This central column is essentially an extension of the turbine tower. Like the tower, it is designed to support the wind turbine, but has additional hydrostatic and dynamic loading. The upper hull, at 7.2 meters in diameter, is the only portion of the TLTP that pierces the waterplane,

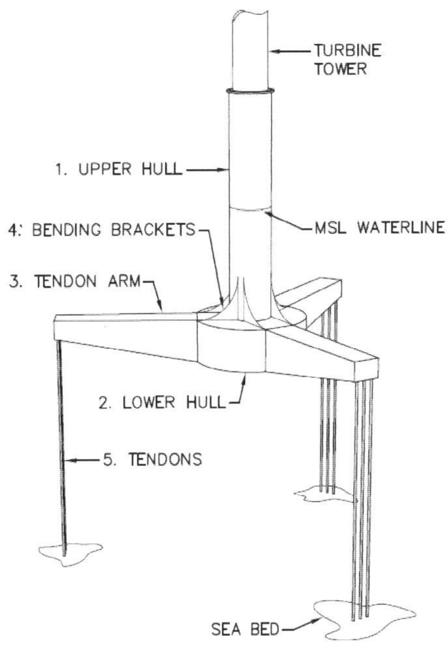

Figure 2: TLTP primary geometry

even during extreme wave events. The mean sea level (MSL) waterline is shown in Figure 2 for reference. The upper hull is made of rolled steel with internal ring frames. Watertight bulkheads are used to subdivide the upper hull to limit flooding in damage conditions.

2. Lower Hull – The primary purpose of this component is to provide sufficient pretension to the TLTP moorings, and to achieve cancelation of the hydrodynamic forces between the various buoyant components. The lower hull is not in the direct load path, but still sees considerable hydrostatic and dynamic pressures. Like the upper hull, it is made of cylindrical rolled steel with internal ring frames. The lower hull is subdivided with vertical watertight bulkheads that are designed as tank boundaries.
3. Tendon Arms – Three tendon arms connect the upper hull to the tendons. The tendon arms experience significant bending loads, hydrostatic and hydrodynamic pressures. They are made of flat plate with internal stiffeners running the length of the arms (i.e., longitudinal). Transverse web frames and cross ties, similar to those in ship tank structures, break up the span of the stiffeners. The tendon arms are buoyant, and subdivided with vertical watertight bulkheads.
4. Bending Brackets – Large external brackets are used to reduce stresses in the upper hull to tendon

arm connection.

5. Tendons – Although unusual for a tension leg platform, the TLTP has tendons comprised of offshore studless chain. Three chains make up each tendon group, for a total of nine tendons per platform[3]. Upper terminations are made with adjustable chain stoppers. Lower terminations to the single anchor chain are made with a custom "tri-plate".
6. Anchors (not shown) – Either driven plate or drilled and grouted anchors can be used to secure the tendon groups to the soil. This choice is largely dependent on water depth and soil conditions.

A preliminary steel weight design margin for small vessels is typically 8%; however, due to the lack of industry experience with wind turbines mounted on tension-leg platforms, a higher 15% margin, consistent with concept design, is used.

The TLTP is designed to withstand damaged chain tendons, in accordance with API-RP-2SK. The TLTP is also designed to withstand single compartment flooding, and has subdivision to permit survival in a reduced extreme environment.

A small amount of fixed water ballast can be incorporated into the lower hull. This allows the pretension to be adjusted and increases the flexibility of the design. As the platform will be unmanned and has no ballast system, a tank level monitoring system is required.

The TLTP presented herein is designed to a combination of guidelines and rules that best suit a first principles "working stress design" approach and a quick design cycle. These include American Petroleum Institute (API) and American Bureau of Shipping (ABS) publications: API RP-2A, API RP-2T, API RP-2SK, ABS Floating Production Installation, ABS MODU and ABS Steel Vessel rules.

Det Norske Veritas (DNV) publishes a set of rules specific to the support structures of offshore wind turbines, but these rules are more onerous (load resistance factor design method) to apply in concept and preliminary design. It is envisioned that DNV or Germanischer Lloyd (GL) rules will be used as the design progresses towards a prototype for a site-specific location. Saigal *et al.* [3] discusses the various differences of the offshore wind turbine design guidelines for United States waters, and Roddier *et al.* [4] discusses standards applicable to floating wind turbines.

DESIGN DRIVERS

At its inception, the design concept required a floating turbine foundation that facilitates complete quayside assembly, thereby precluding costly at-sea turbine assembly and enabling the economical development of wind power in vast stretches of deep offshore areas previously thought to be "un-developable".

Numerous floating wind turbine concepts have been proposed. Spar installation vessel concepts along the lines of the Research Vessel *Flip* [5] have been proposed, but to date have been dismissed by major turbine manufacturers. *HyWind* [6] is a full-scale demonstration of spar technology applied to offshore wind power. *WindFloat* [4], a semi-submersible platform, is presently under development. *Blue H* [7] is developing a tension-leg platform, albeit with a proprietary turbine.

A tension-leg platform, in addition to facilitating complete quayside assembly, has the potential to satisfy intermediate water depths and provides a platform suitable for currently available offshore wind turbines, without requiring major modification. The tension-leg concept is vertically constrained, and thus its heave, roll, and pitch motion are minimal. This fits particularly well with current offshore wind turbines, which lose efficiency as the blades move away from vertical. Minimizing roll and pitch also reduces gyroscopic forces that develop with a pitching, spinning body, such as yaw induced pitch.

Earlier research indicated high potential for tension-leg platforms, but warned that complexity and costs typically associated with their mooring systems were the primary design challenge; i.e., Musial *et al.* [8].

To simplify and reduce the cost of the typical TLP mooring system, chain tendons are used along with driven-plate anchors. For relatively shallow water applications, the simple installation and inexpensive cost of chain far outweighs the small weight penalty. Driven plate anchors are fabricated from widely available flat plate and wide flange beams, and can be designed to have very high holding capacities. They must be hammered and jetted into position, and thus their application water depth is limited. Alternatively, drilled and grouted or suction pile anchors can be installed. Anchor choice is largely a function of water depth, site specific soil conditions and cost.

Each TLTP will have a fully erected and commissioned wind turbine installed while floating at its staging site. The TLTP will be towed to the site and installed, which eliminates the need for costly jack-up installation crane vessels. These vessels will become increasingly expensive at greater water depths.

Because of the large overturning moment of a wind turbine, significant horizontal separation between the tendons is required. This minimizes pitch and roll, while keeping the tendon tensions reasonably low. For this reason, tendon arms are proportionately longer than what is seen on offshore oil and gas TLPs.

Three tendon arms spaced 120 degrees apart provide the minimum number of stabilizing legs to resist overturning moments from any direction. The three-arm system has the added benefit of symmetry around two axes, as the moment of one arm is equal to the moment contribution of the other two arms [1 = 2 x cos(60°)].

Unlike most offshore tension leg platforms, which are "one-off" structures[4], 30 to 100 or more TLTPs will be produced for one wind farm. This scale requires the TLTP to be designed specifically for high-production assembly, with an installation methodology that strives for speed and minimal cost. Steel weight and structural complexity are the primary components to minimize, as they will be the most prominent drivers of the foundation cost.

The overall size of the TLTP is substantial, as shown in Figure 1; however, it is small when compared to even the smallest offshore oil and gas tension leg platform. The TLTP has less than half the displacement, hull weight, and operating depth of the smallest (Morpeth and Allegheny) and the shallowest (Hutton) oil and gas TLPs [9]. This relatively small scale allows for fabrication in smaller and even multiple facilities to meet the installation demand.

To make the TLTP as cost competitive as possible, the largest wind turbine currently available is used in the design to take advantage of economies of scale. When 7- and 10-MW wind turbines are available, the TLTP will likely be modified.

The near-term challenges, and therefore opportunities, in the offshore wind sector center around intermediate depths. For example, sites in the UK Round 3 Development Area primarily have water depths of 35 to 60 meters. There is a gap in cost-effective foundation technology at these water depths. Rather than determining how *deep* a bottom-fixed foundation could go, we set out to determine how *shallow* a floating foundation could go.

The TLTP concept has been shown to be cost competitive when compared with bottom-fixed turbine foundations in water depths of around 50 meters and greater. To become a central solution to the intermediate water depths, the TLTP must operate in as *shallow a water depth* as possible.

DESIGN AND ANALYSIS PROCESS

This section describes the unique, analysis-intensive design process developed for the TLTP concept. Design challenges are identified, a unique design spiral is presented, and design tools are discussed.

Design Challenges

The three most notable design challenges encountered with the TLTP are: 1) determination of the hydrodynamic loads, aero-servo-elastic loads, and their coupling effects; 2) development of a platform mooring system with sufficient pre-tension to balance large overturning moments and prevent slack lines, while remaining sufficiently soft, or compliant, in surge, sway, and yaw; and 3) development of a design process that facilitates solving challenges 1 and 2 in an efficient manner.

The first challenge is inherent with any floating turbine system. The second challenge is unique to the TLTP concept in shallow water. The natural frequencies in surge, sway, and yaw increase (i.e., system becomes stiffer) with increasing pre-tension and with decreasing water depth. For shallow water, a low pre-tension is needed to avoid surge, sway, and yaw resonance with wave excitation forces. However, a high pre-tension is needed to balance the large overturning moments, regardless of water depth.

To address this shallow water challenge, the primary forcing functions on the TLTP can first be de-coupled and analyzed separately. Hydrodynamic loading primarily occurs between 0.06 Hz and 0.2 Hz, or 5 seconds to 17 seconds. The TLTP should have natural periods greater than 17 seconds, at a minimum, in the horizontal-plane motions (surge, sway, and yaw).

Turbine-induced loading occurs at the rotor rotational frequency and its third harmonic, where there is blade-tower interaction similar to a ship's propeller blades rotating past a strut. These frequencies are known as "1P and 3P, respectively. The TLTP natural frequencies should be higher than the 1P and 3P frequencies in the vertical-plane modes (heave, roll, and pitch). Figure 3illustrates the primary forcing frequencies, with the TLTP natural periods superimposed. Interestingly, bottom-fixed towers are generally designed with natural frequencies that "sneak in" between the 1P and 3P frequency ranges, or from 0.22 Hz to 0.30 Hz.

The first design requirement is development of a system that avoids resonance with environmental and turbine-induced excitation. Note that in Figure 3, the yaw natural period is coincident with wave energy. Ensuing analysis confirmed that the yaw mode is

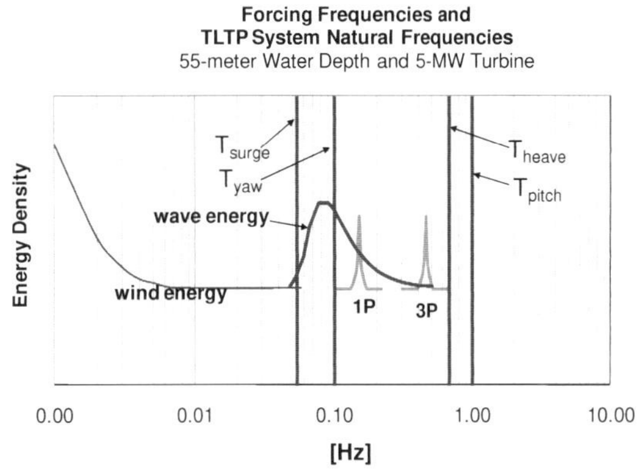

Figure 3: Forcing frequencies and TLTP system natural frequencies.

sufficiently damped by viscous hydrodynamic forces on the arms, and therefore the yaw natural period is not problematic.

The discussion now shifts to the challenges inherent in coupling hydrodynamics, aerodynamics, and control systems. As stated in Nielsen *et al.* [6], the turbine can be an effective damping mechanism against hydrodynamic responses at lower wind speeds, but can lead to "negative damping" when the turbine is in the power-regulated mode (higher wind speeds).

Figure 4 illustrates the relationship between wind speed, rotor thrust, and power. At higher wind speeds, the turbine control system maintains constant RPM to regulate power and avoid excessive tip speeds. When a turbine blade experiences an increase in apparent wind speed, the control system feathers the blade (reduces blade pitch).

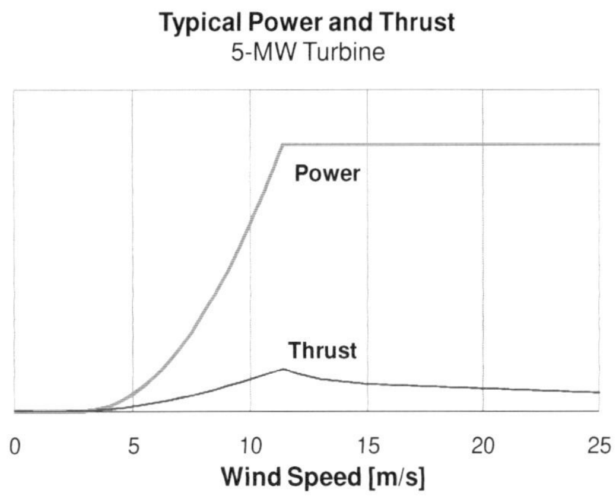

Figure 4: Wind Turbine Thrust and Power, as a function of Wind Speed

Nielsen *et al.* [6] addressed the issue of coupling between surge/pitch motion and blade pitch control by implementing an additional control algorithm aimed at avoiding negative surge damping. This algorithm uses tower velocity as an input.

In a sense, the TLTP faces the same challenge as HyWind [6], though in this case the root cause is pure surge motion as opposed to pitch-induced surge. When the TLTP surges upwind, the apparent wind speed increases slightly and the blades are feathered, which reduces the aerodynamic damping just as it is needed most. Fortunately, the TLTP demonstrates significant hydrodynamic damping in surge. A similar controls-based solution to the surge-blade coupling can be envisioned for this TLTP design as is implemented in HyWind.

Yaw resonance was found to be problematic for a circular barge-type floating wind turbine foundation, especially in the parked condition Jonkman [10]. The TLTP concept demonstrates sufficient hydrodynamic yaw damping to avoid yaw response problems.

DNV OS-J101 gives a comprehensive set of load cases to be analyzed. Jonkman [10] makes a compelling case for analyzing a sub-set of cases including normal operating conditions ("1.X cases"), parked/idling conditions ("2.X"), and extreme sea state conditions ("6.X"). The 6.X load cases are used to optimize the hull form and mooring system based on strength of materials and global platform responses. During optimization, a close eye is kept on the 6-DOF system natural frequencies. Solutions that resonate with normal operating and parked conditions are rejected. The complete set of load cases in DNV OS-J101 will be analyzed in contract design.

Design Process

The TLTP design process is highly simulation-based due to the complex coupling of loads from aerodynamics, control systems, hydrodynamics, and tendons.

A ship design spiral approach simply does not work for the TLTP as it assumes the solution space is smooth. Optimization routines are necessary to find feasible solutions at the beginning of the design cycle, because the solution space is not smooth. Without computer based exploration of the solution space, it is entirely possible (and likely in some cases) that a seemingly good starting point takes the designers along a negative design spiral (spinning outward to an infeasible solution).

A traditional ship design spiral is shown in Figure 5. The first and outermost spiral can be thought of as concept design. This spiral may be repeated many times, until a viable concept that meets all of the design requirements is produced.

Note that in Figure 5 there are no high level

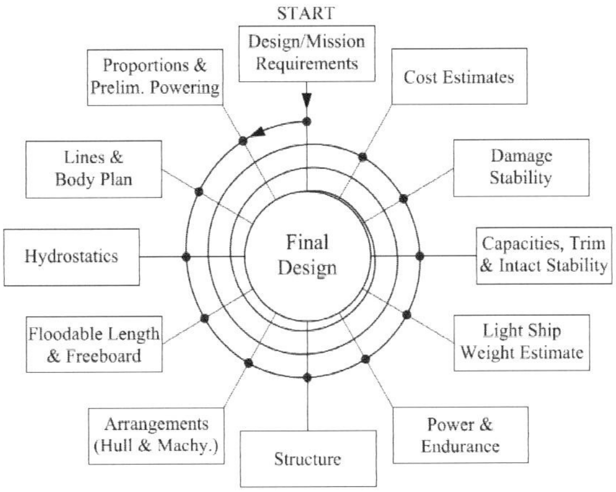

Figure 5: Traditional ship design spiral

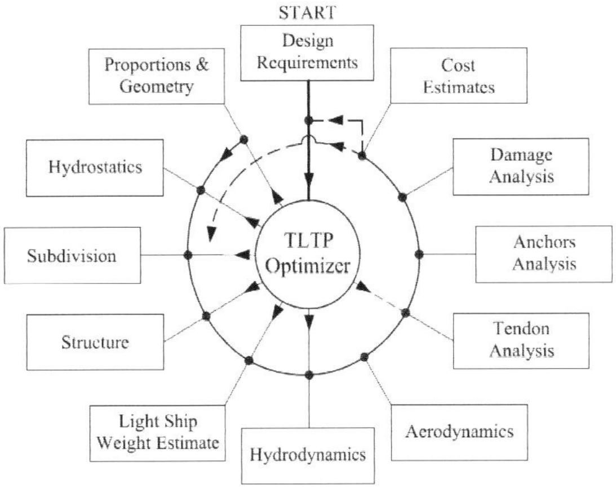

Figure 6: TLTP design spiral

computations required in the concept design phase. As the design progresses from simple to complex (spiraling inward), so do the calculations. Often in the preliminary, and certainly in the contract design phases (inner spirals), hydrodynamic calculations are performed for motions, dynamic positioning, and other investigations such as CFD[5] analysis. The design, however, is typically well detailed at this stage and changes should be relatively minor.

While the authors are not designers of oil and gas TLPs, they observe that the concept design of an oil and gas TLP hull is based significantly on the minimization of hydrodynamic heave forces.

Since the TLTP is designed for relatively shallow water, the relationship between total pre-tension and tendon length can establish a system surge natural period that resonates with wave energy. At the scale of the TLTP, where both inertial and viscous forces govern hydrodynamic loading, there is not an intuitive buoyancy distribution that simultaneously minimizes wave excitation and avoids system resonance with the various forcing functions described above, especially at the target water depths. In other words, it is not sufficient to simply minimize the amplitude of hydrodynamic heave forces at the concept design level for the TLTP.

A unique design spiral (and design tools) was developed for the TLTP concept, which is illustrated in Figure 6. This design spiral is dramatically different from a ship design spiral and more closely resembles the Simulation-Based Design discussed recently in Campana, *et al.* [11]. The TLTP design spiral calls for a significant analysis effort in the early stages of design. Indeed, the determination of a feasible *starting point* requires simultaneous optimization of hydrodynamic loading, system stiffness, and weight.

Similar to the ship design spiral, the TLTP design spiral starts with a clear definition of the design requirements. Moving inward, constraints and objectives are chosen and the optimization is started. The optimizer then radiates the design outward to the outer design elements, whence the design spiral is continued to the last design element: cost. A decision is made at this point to either revisit the design requirements and reenter the optimization, or to proceed with the design spiral. Ideally, solid lines (indicating direct calculations) emanating radially outward from the TLTP Optimizer would be implemented to each design element.

The *art* of design has, therefore, shifted slightly from intuition and experience to creative code writing. This code must contain the flexibility to allow an optimization routine to search, not only unexplored territory, but to perform local searches to achieve an optimal solution.

Design Tools

With the design process established, appropriate design tools must be identified. The present TLTP design was accomplished with a combination of commercial design tools, established research codes, and custom design tools developed specifically for the platform sizing and optimization tasks.

In addition to the typical design tools that naval architects and marine engineers use for ship design, such as 2D drafting, 3D modeling, spreadsheets, word processing, and finite element analysis tools, the following tools are specific to the TLTP design.

Hydrodynamic and Weight Optimization is completed using an in-house tool developed specifically for the TLTP. As with any optimization problem, the objective function and constraints must be well defined. Furthermore, the ability to evaluate the objective in a closed-form manner significantly improves the computational efficiency of the optimization routine. To that end, a simplified model of the TLTP was developed that captures excitation and responses.

The structural weight calculation uses a "rules" based approach, along with first principle combined loads, structural natural frequencies, and simple fatigue, to produce structural scantlings. Intelligent decisions about subdivision are made and weight is calculated.

The optimization has two objectives: minimize structural steel weight and minimize extreme tendon loads.

Constraints on the optimization include: maintain sufficient tendon pre-tension (i.e., do not allow slack tendons), avoid system resonance with excitation forces, and maintain reasonable structural proportions.

With a well defined objective function, well defined constraints, and a closed-from evaluation of the objective function, an evolutionary solver was applied to the optimization problem. Since the evaluation of the objective function involves finite depth wave kinematics and structural step functions, the optimization problem is linear but not smooth.

With an optimized solution that satisfied the design requirements, the feasibility of the solution was verified using a more robust analytical tool.

HARP [12] is a fully coupled, 6-DOF, non-linear, time-domain software capable of modeling the TLTP and its mooring system. *HARP* was used to validate our simplified 3-DOF model and to verify the validity of the optimized solution. For all its strengths, *HARP* is limited to modeling wind loads on "passive" superstructures and rigid bodies. The complex aerodynamics, control systems, and structural dynamics of a wind turbine are not modeled in *HARP*.

FAST [13] is an aero-servo-elastic wind turbine analysis code that captures the lifting surface aerodynamics, control systems, and structural dynamics of a wind turbine. Jonkman [10] made a major break-through in developing a hydrodynamic and mooring module for *FAST*, known as *HyrdoDyn* [10]. HydroDyn captures hydrodynamic radiation and diffraction responses, as well as viscous effects on the submerged section of the turbine tower. Even so, *FAST* with *HydroDyn* requires further modification to model the TLTP, since horizontal-plane Morison members (needed to model the TLTP arms) cannot be implemented in the current version of *HydroDyn*.

CONCLUSIONS

The tension-leg turbine platform as presented in this paper is a feasible foundation for utility-scale offshore wind turbines sited in water depths of 40 meters and greater. The concept can be modified to suit a wide range of turbine sizes, water depths, and metocean conditions.

The TLTP design spiral presented herein lays out a path for developing an optimal solution where the interaction of hydrodynamics, weight, and natural periods dominate the solution space, and where a traditional design spiral fails.

Although *HARP* and *FAST* both have many strong points, neither is the "one-stop" software solution to the design challenges presented here. The *Hydrodynamic and Weight Optimization* would ideally incorporate either *FAST* or *HARP* in the optimization loop to compute the hydrodynamics. A more accurate structural and weight calculation can be incorporated as the structural design is progressed.

Additional analytical modeling and physical modeling are needed to better understand the TLTP system responses, especially during the power production mode. Further development of the TLTP will benefit from the experience and lessons learned from the HyWind project [6], especially with respect to physical model testing and wind turbine control systems.

While initial design phases require optimization of environmental loads and steel weight, ensuing design phases should include optimization based on cost and power production, which are the real design drivers of wind farms.

ACKNOWLEDGEMENTS

The authors would like to thank Carbon Trust and the Offshore Wind Accelerator partners, as well as The Glosten Associates, for funding and support of the work described herein. Our supporting design team includes Bittner-Shen Consulting Engineers, John Halkyard and Associates, and DNV Global Energy Concepts. In kind support was provided by the National Renewable Energy Laboratory.

NOTES

1. Deep water for offshore wind turbines is greater than 50 m. See Pearce et al. [1] for a discussion of shallow water (0 to 30 meters) foundation technology.
2. Det Norske Veritas, Design of Offshore Wind Turbine Structures
3. Cost of end terminations may drive the number of tendons to be decreased.

4. TLPs Morpeth and Allegheny, as well as Kizomba A and Kizomba B, are the only TLPs that claim "sister" status.
5. Computational Fluid Dynamics

REFERENCES

[1] Pearce, K., Geiger, S., and Nordstrom, C. (2009). "State of the Art in Offshore Wind Energy Projects." SNAME Offshore Wind Panel.

[2] Jonkman, J., Butterfield, S., Musial, W., and Scott, G. (2006). "Definition of a 5-MW Reference Wind Turbine for Offshore System Development." National Renewable Energy Laboratory, Technical Report NREL/TP-500-38060.

[3] Saigal, R.K., Dolan, D., Kiureghian, A.D., Camp, T., and Smith, C.E. (2007). "Comparison of Design Guidelines for Offshore Wind Energy Systems." *38th Annual Offshore Technology Conference*, OTC 18984.

[4] Roddier, D., Cermelli, C., and Weinstein, A. (2009). "Windfloat: A Floating Foundation for Offshore Wind Turbines Part I: Design Basis and Qualification Process." *28th Int. Conference on Oceans, Offshore and Arctic Engineering*, OMAE2009-79229, Honolulu, Hawaii, USA.

[5] WindFlip (2010). Web page as downloaded January 2010 at www.windflip.com.

[6] Nielsen, F.G., Hanson, T.D., and Skaare, B. (2006). "Integrated Dynamic Analysis of Floating Offshore Wind Turbines." 2006 European Wind Energy Association Conference, Conference Paper 0749.

[7] Blue H USA, (2010). Web page as downloaded January 2010 at www.bluehusa.com.

[8] Musial, W., Butterfield, S., and Boone, A. (2003). "Feasibility of Floating Platform Systems for Wind Turbines." National Renewable Energy Laboratory, Conference Paper NREL/CP-500-34874.

[9] Mustang Engineering (2007). "2007 Offshore Worldwide Survey of TLPs, TLWPs." Offshore Magazine.

[10] Jonkman, J. (2007). "Dynamics Modeling and Loads Analysis of an Offshore Floating Wind Turbine." National Renewable Energy Laboratory, Technical Report NREL/TP-500-41958.

[11] Campana, E.F., Peri, D., Tahara, Y. Kandasamy, M., and Stern, F. (2009). "Numerical Optimization Methods for Ship Hydrodynamic Design." *2009 SNAME Annual Meeting*, vol 1, Rhode Island, USA.

[12] Hull and Riser/Mooring Program (2010). Web page as downloaded January 2010 at www.harponline.com.

[13] Jonkman, J., and Buhl Jr., M.L. (2005). "FAST User's Guide." National Renewable Energy Laboratory, Technical Report NREL/EL-500-38230.

(This page intentionally left blank.)

SESSION B PAPERS

(This page intentionally left blank.)

THE RHO-CEE WAVE ENERGY CONVERTER - IMPEDANCE-MATCHED TERMINATOR

Neal A. Brown
Float Inc

ABSTRACT

The "Rho-Cee" (ρC) Wave Energy Converter (WEC) system is subtitled "The Impedance-Matched Terminator". It is a large, floating Oscillating Water Column (OWC) system, to be moored in deep water. The name "Rho-Cee" derives from the expression for the characteristic impedance of water gravity waves; the product of water mass density, ρ, with the length-dependent velocity of such waves, C. It is the base principle of the WEC design that its input impedance matches the characteristic impedance of the targeted waves. Impedance matching maximizes the capture of wave energy; with minimum reflection. The original means for extracting power is via unidirectional reaction turbines, driving alternators. Oscillating air flow is between the OWCs and atmosphere; rectified at the turbines. This is more efficient than the commonly used, bi-directional "Wells turbine".

INTRODUCTION

The "Rho-Cee" is integrated with the Float Inc Pneumatically Stabilized Platform (PSP) to take advantage of the controllable stability, load capacity and deck area that it provides.

The Pneumatically Stabilized Platform

The Pneumatically Stabilized Platform (PSP) technology enables the operation of very large floating platforms, moored offshore, with superior motion stability and load capacity in challenging sea-state environments. (Float Incorporated, San Diego)

It achieves its at-sea motion stability and structural loads mitigation by decoupling the "hull" from ocean wave pressures through the use of air buoyancy, which is both compressible and mobile. The air is contained in an array of interconnected, open-bottomed, cylindrical tanks. The air is made mobile by means of ducting. The ducting arrangement is selectable in real time to best suit the sea-state environment and platform deck loading distribution. Local air pressure adjustments are made via Roots-type blowers, as needed.

The PSP is constructed, modularly, in pre-stressed, reinforced concrete, which, if properly formulated and applied, has been shown to be degradation-free in long term exposure to sea water. In contrast to steel hulls, a concrete PSP need never be dry-docked for inspection, and requires no significant maintenance of the basic hull. Its useful life is expected to exceed 50 years.

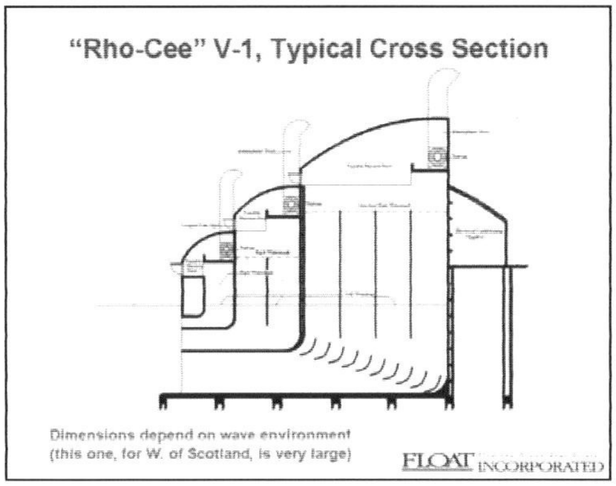

Figure 1 "Rho-Cee" V-1 Typical cross section

An elongated PSP forms the base for a comprehensive renewable energy system. In addition to the "Rho-Cee" WEC, described below, a number of large wind turbines can be supported, with favorable maintenance possibilities – as can a number of deployable / retractable ocean current turbines. All can contribute to energy storage within the PSP to overcome the intermittency of each. The multiple users, including non-energy, marine real-estate users, can greatly reduce the individual capital and operating costs associated with their mutually supporting offshore platform..

The "Rho-Cee" WEC

The "ρc" Wave Energy System is:

- An Impedanced-Matched Terminator
- Constructed in reinforced concrete

Several constraints dictate that the input impedances of the absorber elements be quite small. This requires resonant operation of the OWCs. Hence, several water columns are tuned to frequencies, with band-widths that span the energetic region of the yearly average incident wave power spectrum. The normalized bandwidths govern both the resistive input impedance and the output power potential of each oscillator, see Figure 2.

The successively-tuned water columns are geometrically "nested" to minimize space and weight of materials – hence cost. The nested units are then repeated, endwise, to form a linear array of identical contiguous WECs in a two-dimensional "terminator" configuration – one that is aligned perpendicular to the usual propagation direction of incident waves.

Comparative Advantages

The successively-tuned water columns are geometrically "nested" to minimize space and weight of materials – hence cost. The nested units are then repeated, endwise, to form a linear array of identical contiguous WECs in a two-dimensional "terminator" configuration – one that is aligned perpendicular to the usual propagation direction of incident waves.

All wave energy absorption takes place along the face length of the "Rho-Cee" WEC. This is in marked contrast to "attenuators" that are aligned parallel to wave propagation, and to the various fields or "farms" of "point-absorbers". Both inherently suffer from progressively decreasing wave energy availability, front-to-rear. Their elemental capital and (submerged equipment) maintenance costs, however, are independent of front-to-rear position. The "Rho-Cee" is all "up-front", and in deep water. It is expected to deliver 50% of incident power; that is, a "wave-to-wire" efficiency twice that of any other extant system.

The "ρc" Wave Energy System has:

- All equipment in air; none in water
- All equipment accessible for maintenance by personnel on foot

Most significant from a practical (that is economic) point-of-view, is that only concrete touches sea water in the "Rho-Cee" WEC. All equipment subject to maintenance, replacement or inspection is "in-the-dry" – fully accessible to personnel on foot, dry-shod. Maintenance personnel can comfortably live on the PSP platform.

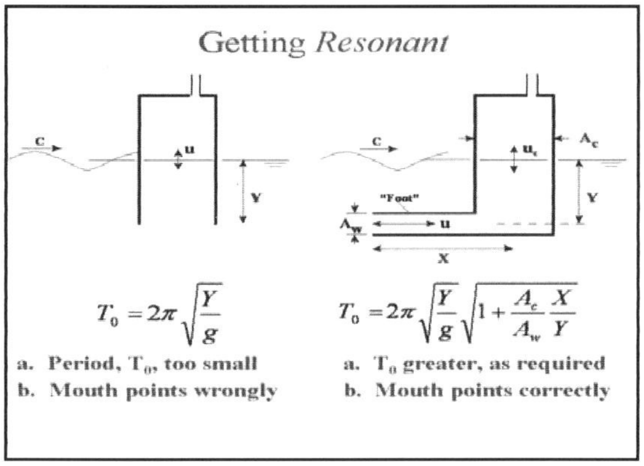

Figure 2 Method to achieve resonance

The original means for extracting power is via unidirectional reaction turbines, driving alternators. Oscillating air flow is between the OWCs and atmosphere; rectified at the turbines. This is more efficient than the commonly used, bi-directional "Wells turbine". We are now examining a potentially more efficient and economical process involving long-stroke hydraulic cylinders with accumulators, hydraulic motors and controls (all COTS equipment). More direct, electro-mechanical take-offs are also on the table. Solid-state power conditioning is a given.

Economy of Power Production is achieved by:

- Maximum wave-to-wire efficiency, beginning with near-total wave-energy capture
- Reduced O&M costs due to ease of maintenance and relatively benign equipment environment
- Small annual amortization cost due to very long useful life: > 50 years
- Ancillary uses of the associated floating real estate

Finally, the "Rho-Cee" WEC is uniquely capable of storing substantial amounts of potential energy in its supporting PSP structure. That energy can be tapped during intervals of low wave activity to better match varying demands of the load infrastructure, thereby avoiding the principle objection to renewable energy sources. In a "primitive" storage scheme, Figure 3, the platform is raised as air is injected into both buoyancy and interstitial volumes – resulting in typical capacities of about 1.5 MW-hr per hectare of planform. If water ballast is applied to maintain draft, and closed interstitial volumes are raised to three atmospheres by blowers staged in series, energy storage capacities of about 10 MW-hr per hectare are possible. Efficiency and practical considerations suggest that heat be added to the power producing expansion process – either by the combustion of a volatile fuel or, if staged, by the use of relatively warm sea water.

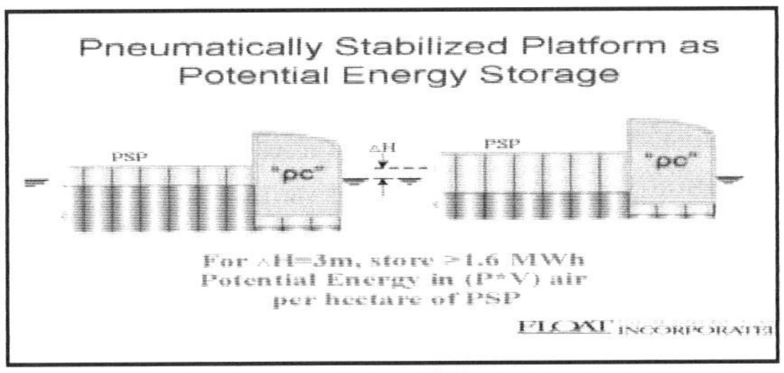

Figure 3 PSP as potential energy storage

Experimental Program Objectives

To verify the impedance-matching basis of the Rho-Cee WEC concept and design a model test program was prepared. This could potentially reveal any discrepancies between performance expectations and observations so that they may be diagnosed, remedied and be subjected to further experimental verification.

Finally the tests provide a graphic demonstration of Rho-Cee WEC performance and that of its associated Pneumatically Stabilized Platform (PSP) in the wave environment(s) of its intended installation(s) – yielding presentation materials suitable for promotion of the concept program.

The study was carried out in two phases. In the first phase the tests were limited to two-Dimensional Models, in the second phase a full three dimensional model was used.

THE FACILITY

The U.S. Minerals Management Service (MMS) is collaborating with research institutions and industry on testing wave and current energy systems. Ohmsett, The National Oil Spill Response Research & Renewable Energy Test Facility, is one of the largest outdoor saltwater tanks in North America and is designed to evaluate the performance of full-scale and meso-scale equipment under realistic but safe environmental conditions.

The facility, managed by MMS and operated by Mar, Incorporated, is located at the Naval Weapons Station Earle Waterfront in Leonardo, New Jersey (about one hour drive south of New York City).

At the heart of Ohmsett is a large outdoor, above ground concrete test tank that is 203 meters long, 20 meters wide 3.5 meters (8 feet) deep and holds 10 million liters of crystal clear salt water. The Ohmsett tank is equipped with three movable bridges with tow speeds of up to 6 knots, programmable in 1/100th knot increments to simulate ocean current flow. The robust tow bridges are able to accommodate the torques and forces of the largest current turbines and wave energy converter (WEC) equipment. The Control Tower is fully computerized and data from various sensors and video cameras are collected for synthesis and analysis.

The Ohmsett tank system allows testing of full-scale equipment. The tank's wave generator creates realistic sea environments, while state-of-the-art data collection and video systems record test results. The facility has proven to be ideal for testing equipment, evaluating acquisition options, and validating research findings.

The wave generator system can produce waves of discrete frequencies with heights up to one meter, FM Slides, JONSWAP and Pierson-Moskowitz spectra with controlling parameters of driving wind velocity and scale ratio. Recognizing physical limits on wave-maker stroke, probabilities of clipping on axes of these two parameter values are shown in Figure 4 for the P-M spectra. Figure 5 shows the theoretical stroke-to-waveheight transfer gain of the wave flap – with a number of measurement observations superimposed. Customers may specify their own wave characteristics if they are not adequately represented by these standard types. The data collection system is capable of recording up to 32 channels.

The facility includes a well-equipped machine shop and extensive lifting and transport capabilities. Figure 6 shows the four-bay "Rho-Cee" WEC model, which with ballasted base, weighs more than two tons, being installed in the tank. A new North end beach system is being developed that will allow long runs of variable wave incidence without interference from reflections.

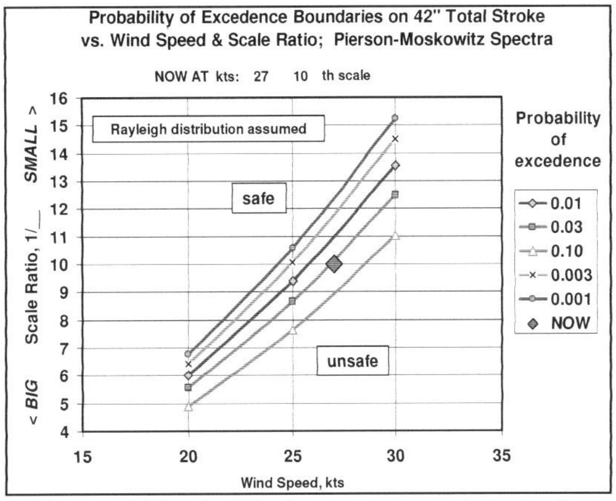

Figure 4 Physical limits on wave maker stroke

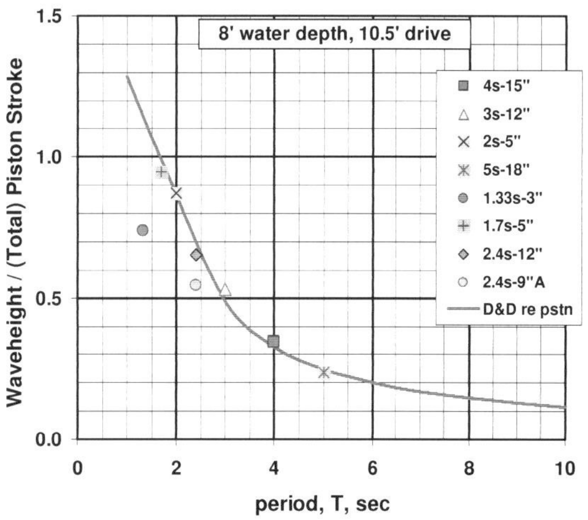

Figure 5 Transfer gain for full piston stroke

Figure 6 'Rho-Cee' WEC model being installed in the tank

MODEL

Design basis

The principle of impedance matching, when applied to a Wave Energy Converter (WEC) energy capture, can be expressed in design terms by the following:

When forced by the doubled pressure of a blocked incident sinusoidal wave, the submerged mouth of the absorber must sinusoidally emit a quantity of water that will generate a wave of the same frequency with the same amplitude as that of the unblocked incident wave in opposite (180 degree) phase.

This will be possible if the WEC oscillator is operating at resonance, thereby presenting a small, resistive input impedance. In principle, that resistance will be controllably applied by the power take-off mechanism whereby the oscillations are converted to useful work. If done as stated, the generated wave will remove the doubling of the blocked incident wave and render the WEC-containing baffle "invisible" to the incident wave. There will then be no reflection of wave energy from the baffle; all of the incident wave's energy will be absorbed by the WEC.

For deep water, the wave-making transfer function can be expressed, per Newman (4), when discretized as:

$$a \approx \frac{2\omega^2}{g} y_C A_C \exp(-kY)$$

and in shallow water, similarly, per Dean & Dalrymple (5), as:

$$a \approx \frac{\omega^2}{k\,g}\left(\frac{A_C}{A_W}\right) y_C \Big/ \cosh\big(k(h-Y)\big)$$

where: **a** is the amplitude of the generated wave (same as that incident), **ω** is the radian frequency, **g** is the gravitational constant, $\mathbf{y_c}$ is the amplitude of vertical motion of the water surface in the column, $\mathbf{A_c}$ is the cross-sectional area of the column (and the water surface there), $\mathbf{A_w}$ is the area of the submerged horizontally directed mouth, **Y** is the mean depth of the area Aw, (**X** being the horizontal leg mean length), **k** is the wavenumber corresponding to the frequency and the depth of water, and **h** is that water depth.

Note that the quantity (ω yc A_c) represents, by continuity, the amplitude of the volume rate of flow into and out of the submerged mouth of the WEC.

Now the ratio, a / y_c , is directly related to the loss factor, η , which happens to be the inverse of the resonant response amplification factor, commonly referred to as "q". More important, η is also equal to the normalized bandwidth of that resonant response, be it expressed in either frequency or period. (See "Some Test Results" for more on η.)

The designer may now lay out the geometric arrangement of a Rho-Cee (ρC) WEC subject to any number of constraints and goals. One of these is that the upper edge of the submerged mouth, Aw, be deep enough to avoid venting in the highest waves expected while producing power in resonance. There is no codified direct procedure for getting to a satisfactory result; it is iterative and intuitive.

The Design for Ohmsett

We would usually span the economical frequency range in three adjacent parts, or resonant columns. However, for application at Ohmsett where the depth is limited to about eight feet, and where the un-subsidized cost of materials and construction must be considered, it was decided to include but two nested columns, thereby limiting model draft, handling issues and cost of materials. Proof-of-principle was considered preserved with but two adjacent bands. The cross-sectional layout is shown in Figure 7; dimensions in feet, the waterline-front at (0,0). The resonant wave periods of 2.12 and 2.83 seconds were chosen with normalized bandwidths of 34.83% and 23.16%, respectively. (The latter being one-third of an octave in width; a common BW used in acoustic and vibration spectral analysis.) The longer period column was located at the rear and denoted as "back" and the shorter period forward, denoted "front". Freeboard of 18" was added above the waterline as more than adequate to contain resonantly amplified oscillations of the column water.

Four identical 4-foot modules were constructed and connected endwise to form a 16' long array of WECs. When placed in the Ohmsett tank with one end against the West wall, the effective array length became 32 feet. While not quite two-dimensional, analysis of measured data was concentrated on the "B" module; removed by but one module-length (that of the "A" module) from the wall at center length.

The emphasis on four feet was dictated by plywood panel width. Vertical and horizontal column partitions were made up of three layers of nominally ¾" plywood – although the true thicknesses were found to be somewhat smaller. The resulting less-than-the-planned 2.25" thicknesses proved troublesome; requiring some adjustments. End plate partitions were each a sandwich of two ½" plywood panels; the interior layer being cut away to form rabbets for the accurate and rigid placement of the longitudinal partitions and turning vane end plates. After assembly, the noses of

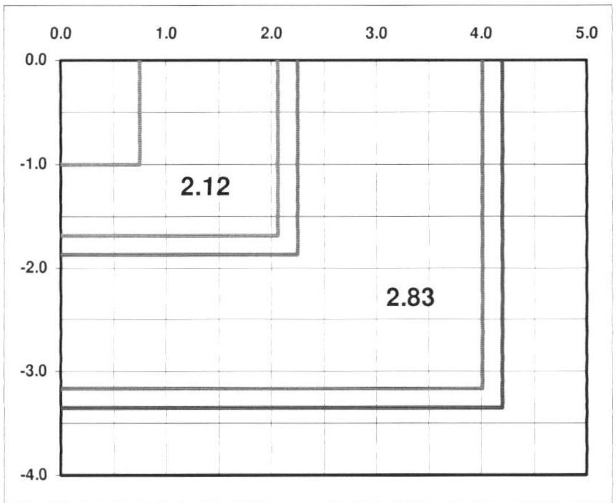

Figure 7 Cross-sectional layout

the module partitions, now 2" thick, and the upper and lower port edges were fitted with half-round moldings to fair the openings and, hopefully, avoid some entrance losses.

A steel base with angle-bar truss-work was welded up, to which the modules were bolted. The height of this assembly was intended to support the WEC at its design waterline when the base rested on the flat tank bottom.

The spreadsheet calculations in Figure 8 detail the dimensions of the Ohmsett Rho-Cee WEC. They are set up to allow manipulation of certain dimensions and parameter values while others are held fixed. The output of the calculation is the resonant period, To, in the 8th column that must be made to agree with that value entered in the 1st column (both green). The bandwidth, BW, which is the same as the loss factor, η, is to be noted; colored tan.

									T bands
		BW:	0.3483	Y1:	1.00	X1:	0.75		1.782
2.12	sec	k:	0.280	omega:	2.965		g*eta/2omega^2	0.637	2.119
				0.9774	0.280				2.520
Aw		Y	Ac	X	X/Y	kY	Ac/Aw	To	
0.69		1.300	1.318	1.245	0.958	0.363	1.911	2.12	
		BW:	0.2316	Y1:	1.878	X1:	2.26		2.520
2.83	sec	k:	0.174	omega:	2.221		g*eta/2omega^2	0.755	2.828
				0.8831	0.174				3.175
Aw		Y	Ac	X	X/Y	kY	Ac/Aw	To	
1.29		2.439	1.761	3.010	1.234	0.424	1.365	2.83	

Figure 8 Rho-Cee for Ohmsett, V4 2.25" partitions

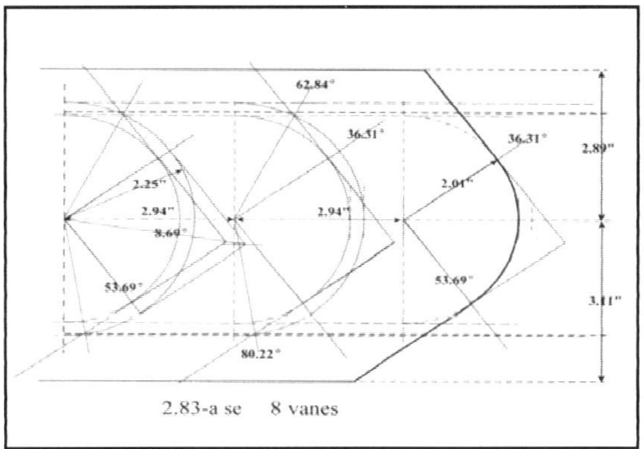

Figure 9 Layout of the turning vanes

Turning Vanes

The vanes are intended to be simply sawn out of 4"schedule 40 plastic pipe. The two saw cuts required are in planes at 90 degrees to one another and terminate in tangents to the interior surface of the pipe. The symmetric, sharp-edged section created thereby subtends 143 degrees, measured from its (the pipe's) centerline. While all the vanes had identical sections, their gaps and setting angles were different in the two column's cascades. Based on experience with potential theory cascade design, the gaps and angles were adjusted to produce an "over turning" of about 4 degrees in both flow directions – anticipating a real-flow performance that would be compensated. Edges were blunted by running them through a jointer and subsequent sanding. Figure 9 shows the layout of one end of a ½" plywood end plate that, in terminal pairs, regulate the spacing and angle geometry of the cascade. The particular detail is that of the "South-East" end of the 2.83 second cascade; the vanes, sections in color, let into cut-outs in the 6" wide end plates. As the (slightly less than) 47" long vanes were quite flexible, they were regulated and supported by three tie-rods of ¼"-20 all-thread in stainless steel. The vanes were located by pairs of low-profile lock nuts, while the tie rods butted against the fixed quarter-round pipe sections at the inner and outer cascade-terminating "elbows" of the column partitions. Labor and materials for finishing and assembling the cascades were generously provided by Ohmsett technical staff – and constituted a major part of the effort to assemble the model WECs.

On Linearity

The principle of impedance matching, employed to maximize the capture of energy from the incident waves, requires that the input impedance presented by the face of the WEC be resistive and linear. That is, in the water at the mouth of the OWC, the oscillating pressure should be in phase with, and proportional to the water velocity there – at frequencies equal to and near the resonant frequency of the OWC. Failing this linearity, reflections from the WEC face are to be expected which represent un-captured, lost energy.

Ideally, the full resistive load in the OWC is provided by the power take-off. Depending on the nature of the power equipment, its operation must be continuously controlled to maintain the desired linearity. Thus an air turbine requires that its inlet nozzle area be controlled to maintain constant the ratio of air pressure-to-weight flow rate; a hydraulic motor, its displacement controlled to maintain the ratio of hydraulic pressure-to-flow rate (proportional to piston velocity); and for a linear generator, its armature current controlled to maintain the ratio of field rod force-to-velocity. Un-controlled, such "pressures" would naturally vary as the square of their related "velocities". This would result in the generation of frequency harmonic content ("ripple") in the presented impedance.

Since the hydrodynamic flow losses in the OWC and in the air or hydraulic fluid power flows also vary as the square, they may best be compensated-for by controlling on the basis of water oscillating pressures and velocities sensed directly at the mouth of the OWC. Sensors for this duty may be positioned there, and, as a practical matter, must be retractable for calibration and replacement.

TESTING, AIR VALVES

Abandoning the quest for an elusive linear air flow resistance element, an easily adjustable "square-law" device was sought and found in the form of a "blast-gate" variable area slide valve. These are simple gate valves, used in sawdust vacuum systems in woodshops. They are manufactured in plastic and are quite inexpensive. A satisfactory unit of 6" nominal size was available from McMaster-Carr. Ohmsett ordered fourteen of them – which were delivered within hours. Two valves were to be fitted to each of the larger 2.83 second columns and one to each of the smaller 2.12 second columns.

A calibration rig was quickly assembled by Ohmsett staff in the form of an ASME-type orifice air flow meter followed by an "air-box" that discharged through the valve under test. Air was supplied to the flow meter entrance piping by a gasoline-powered leaf blower. The air box was fashioned from a steel barrel with the subject valve led into its lid. A ¾" thick plywood ring plate was bolted to the lid – into which the valve was inserted – in order to simulate the valve's installation in the water column's cover plate. Flow meter orifice differential pressures were measured on a U-tube water manometer, while the air box – to – atmosphere pressure drops across the valve were measured by a 3"WC Dwyer slant-tube manometer.

The valve's slide position was calculated for nominal airflow area in 2 in^2. Increments and the slide valve was so marked for easy adjustment. Using the airflow rates available from the leaf blower, the pressures were measured and the valve's pressure drop coefficient calculated for each geometric setting of the slide.

Pressures were measured first with the valve outside the air-box lid – as in the "exhale" mode of the WEC oscillation and, by inverting the lid, so that the valve was inboard, as in the "inhale" mode. It was found that the Cp's on "inhale" were considerably greater than on "exhale", presumably due to the sharp edges of the 6" "Borda" inlet with that flow direction. A 12" OD ring of ¾" ply was applied to provide a limited "flush-mount" and the measurements repeated. The results showed a more symmetrical set of pressure coefficients, while still different, and the rings were applied to the outside of the valves installed on the Rho-Cee WEC model covers.

The valve flow areas were fitted to the Cp's in a

second order polynomial of their logarithms and used to calculate valve settings vs. flow rates for design values of column loss factors and for several lesser values, assuming internal losses were present (which they were). From these, a map of test parameter values was laid out which was subsequently partially covered in the available test time.

SOME TEST RESULTS

Some sample test results are presented for 60 seconds duration in the "strip-charts", produced in Microsoft Excel. These are for Test 03, with a period of 2.83 seconds and a commanded wave-flap drive piston amplitude of 7.8" resulting in an incident wave amplitude of approximately 4.3". In all cases, attention is directed to the "B" module of the four-module WEC; that being the second from the tank wall. In this case, the data is for the "Bback" column that is tuned for the 2.83 second resonance.

Figure 10 shows the column air pressure, sampled at 0.1 second intervals, is plotted against time, showing a clearly non-linear characteristic that reflects the square-law pressure resulting from the oscillating airflow through the pair of slide valves. In this case, each valve is set to an opening of 10 in^2. To determine the corresponding airflow history, the pressure coefficient appropriate to either "inhale" or "exhale" is applied to either the negative or positive pressure values, respectively. The resulting corresponding airflow history is shown in Figure 11. Surprisingly, the airflow shows very little, if any, of the non-linear, square-law behavior apparent in the pressures. The airflow is quite sinusoidal in its variation. In fact, a spectral analysis of the airflow "record" shows that the harmonic content of the signal is very small in comparison to the fundamental. In this case, the twice frequency component has about 1/20th the amplitude of the fundamental representing, here, about 1/400th of the fundamental's energy (~ 0.25%). Higher harmonics are yet smaller. (Spectral analysis is performed by employing a 512 point Fast Fourier transform, in MATLAB, see Figure 12) At this point it is convenient to directly estimate the power dissipated by the pressurized airflow by simply multiplying the pressure and airflow records, in compatible units, and averaging the result, time-wise. In this case the air power is estimated at 74.5 Watts (quite typical).

The nearly pure sinusoidal nature of the airflow history has important implications for the practical development of the Rho-Cee WEC – as will be discussed in the Conclusions section.

With the amplitude of the air flow-rate in hand,

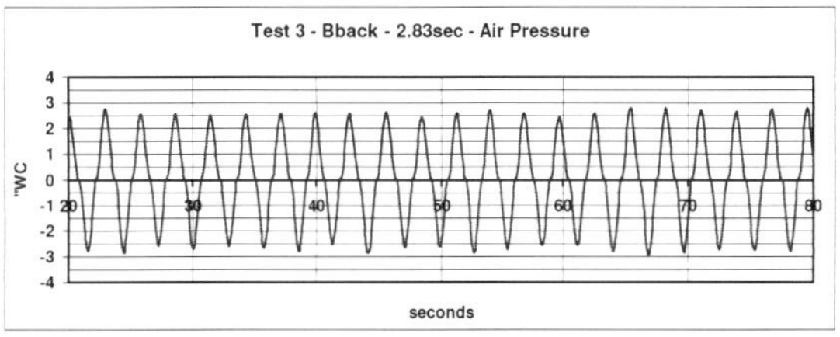

Figure 10 Test results for air pressure

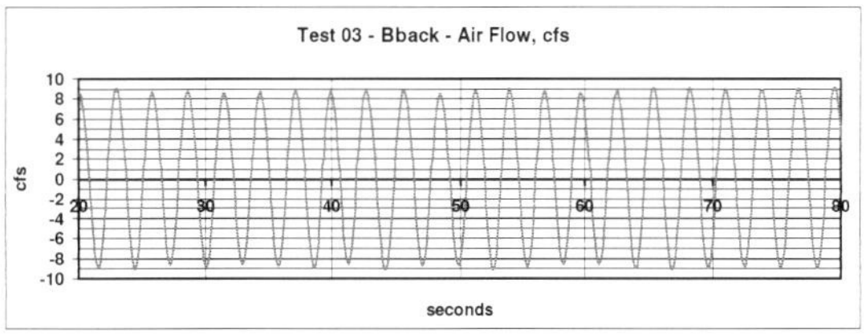

Figure 11 Test results for air flow

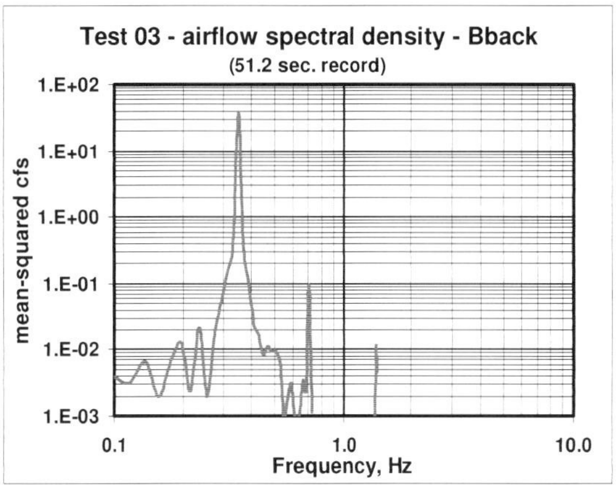

Figure 12 Air flow spectral density

and knowing the surface area of the column, it is easy to estimate the velocity and amplitude of the column's water surface average motion, yc. With the wave amplitude, a, and channel mouth depth, Y, it is now possible to estimate the overall loss factor, η, of the process – and to compare the result with the desired band width, whose normalized value it represents.

$$\eta = 2\frac{a}{y_c}e^{-kY}$$

where k is the wave number, in deepwater. Similarly, that part of the loss factor that is manifest in the power extraction of the throttling air system, ηa, can be estimated by invoking the relationship between that power, Pa, the kinetic or potential energy of the oscillating water system, E, and the radian frequency, ω:

$$P_a = \omega\,\eta_a\,E$$

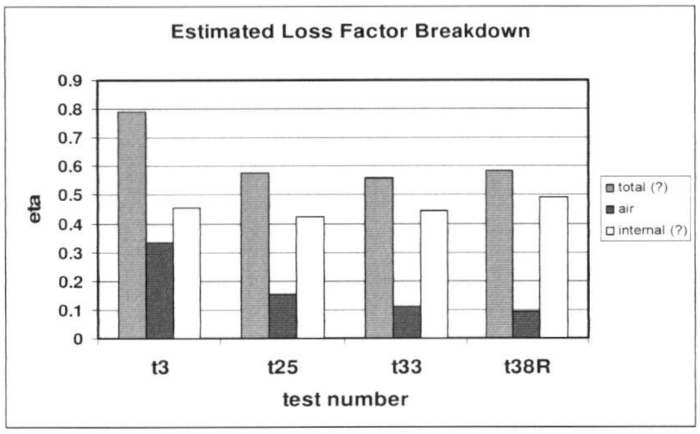

Figure 13 Estimated loss factor breakdown

where the energy is directly related to the column area, Ac, and the amplitude, yc , squared.

An estimate of the internal, parasitic loss factor is obtained by subtracting the calculated air-power loss factor from the estimated total loss factor. The results for four tests on the "Bback" column with resonant, 2.83 second waves are shown. Recalling that the intended total loss factor, to be mostly in the air-power, was 0.2316 (equivalent to a 1/3rd octave band), it is clear that the internal losses are far greater than that. It is also apparent that the internal losses are quite consistent, for this set of four samples, with an average of 0.454 ±0.024, in face of much larger variations of both the total and air-power components. Examination of the photo (Figure 14) of the water surface of the (in this case, "front") column, below, shows intense turbulence, strongly suggests that the turning vanes are responsible. While highly efficient in steady flow, a cascade of turning vanes becomes a source of strong shed vorticity when operating in an oscillating flow. The vane cascades become dampers.

SOME PRELIMINARY CONCLUSIONS

Maintenance of loading linearity is over-stated.

This consideration, the controlled maintenance of a constant value of the input impedance throughout the wave cycle, was considered to be very important in eliminating reflection losses at harmonic frequencies. However, while air pressures show the expected square-law variation with the air flow rate, the air flow rates themselves are found to be very sinusoidal in shape. The worst case found so far (limited sampling of results) shows a 2x frequency component of less than 4% of the fundamental on an energy basis. As the air flow rate (compressibility ignored: pressures only 3"WC in 33' atmospheric) is a direct measure of water flow rate, hence generation of the reflection-canceling wave, it appears that a relatively negligible non-linearity is presented to the incident wave system. The

Figure 14 Water surface in the front column

absence of a need for linearity control, if it proves true, will greatly simplify and reduce the cost of a practical Rho-Cee WEC system.

The use of a cascade of turning vanes introduces damping.

While turning vanes are highly desirable in steady flow, their use in an oscillating flow (with both magnitude and direction) unavoidably incurs a significant energy loss. They behave as dampers! Even ignoring viscous flow losses, the non-viscous vorticity shedding amounts to a major loss factor value, in a quick estimate. Since a large cost of construction and maintenance of a full scale Rho-Cee WEC may be involved, and 2/3rds of the effort of building the model, their elimination may be a large benefit. We will revert to a long- radius bend approach for further efforts.

REFERENCES

[1] "Float Incorporated and the "Rho-Cee" WEC," White Paper in response to Minerals Management Service RFP M08PS00094, Float Incorporated, August, 2008.

[2] "A Broad-Band Ocean Wave Power System for the West of Scotland,". Ocean Energy Conference, SNAME CIMARE IMAREST, 12 February 2005, N. Vancouver, B.C. Also: Energy Ocean 2006, 21 June 2006, San Diego, CA.

[3] "Design Approach for a Broad-Band Oscillating Water Column (OWC) Ocean Wave Power System", Float, Incorporated, June 2003

[4] J. N. Newman, "Marine Hydrodynamics", The MIT Press, Cambridge, 1977.

[5] R. G. Dean & Dalrymple, R. A., "Water Wave Mechanics for Engineers and Scientists," World Scientific, Singapore, 1991

SIMULATIONS AND MODEL TESTS ON A WAVE ENERGY CONVERTER BASED ON INVERSE OFFSHORE ENGINEERING

Bas Buchner and Koos Hoefakker
MARIN (Maritime Research Institute Netherlands)

BACKGROUND

For more than 75 years, the Maritime Research Institute Netherlands (MARIN) has been contributing to the development of safe and economic ships and offshore structures as independent advisor. This is done through model tests, simulations, desk studies, full scale measurements, safety studies and training. With the resulting knowledge of the ocean environment and the hydrodynamics of ships and offshore structures, MARIN sees it as its responsibility to contribute to the development of renewable energy offshore from waves, tides and wind:

Wave energy is a concentrated form of wind energy: the wind transfers energy into the waves over a long fetch. Cruz [1] has estimated that Europe has an average wave power of 50kW per metre width of wave front [1]. Others [2] indicate that worldwide the economically exploitable amount of wave energy is estimated at 2,000 TWh/year, an average power of 200GW over a year. This is equivalent of 200 large power stations. The challenge is to generate a predictable amount of energy, in a reliable way, at a reasonable cost.

Figure 1: The 'Green Water Concept' uses knowledge about ship motions to develop a wave energy device through 'inverse engineering'

These challenges of wave energy are very similar to those of the offshore industry: safe and economic design, production, transportation, installation, maintenance, repair and removal. Besides that, knowledge about reducing motions from offshore hydrodynamics can be used by 'inverse engineering' to increase the motions of wave energy converters to maximise the wave energy conversion into useful electrical energy [3-5]. The present paper focuses on the development of a wave energy converter based on these principles.

But in a similar manner, MARIN knowledge in the field of ship propulsion systems can be used to optimise marine turbines to generate tidal energy. The advantage of tidal energy is that the amount of energy is more predictable than wave energy (but its presence is much more localised). MARIN has developed an extensive palette of computational tools and experimental capabilities that suite the design and evaluation of state-of-the-art propellers, see Figure 2. For turbine rotors MARIN has already employed its tools and capabilities with panel and CFD codes for evaluation and design of practical turbine rotors.

Finally, offshore wind energy is much related to normal floating and fixed offshore structures. Installation, removal, maintenance, survivability and

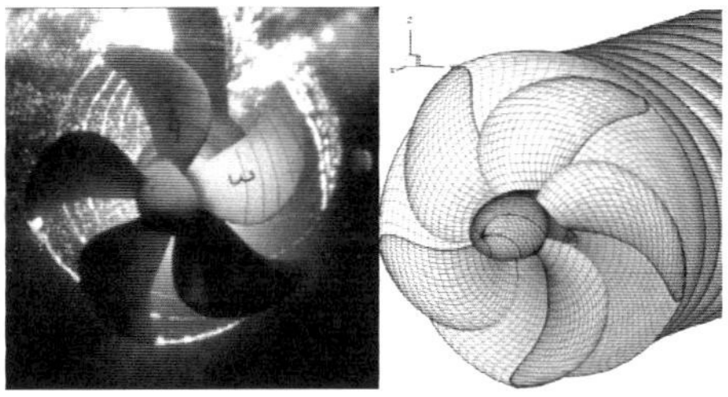

Figure 2: Tests and simulations can be used to simulate both ship propulsors and tidal turbines

vessel traffic safety are topics that link offshore wind energy to MARIN's broad maritime expertise. The Institute also contributed to the development of windmill installation vessels. This included resistance and propulsion of the special hulls, Dynamic Positioning (DP) of the system at the location and the actual installation procedures by simulations and model tests.

GREEN WATER CONCEPT DEVELOPMENT

To use its ship and offshore hydrodynamic expertise for the development of renewable energy, MARIN started the development of the 'Inverse Concept' or 'Green Water Concept' for wave energy conversion [5]. The concept is a combination of 'heaving' or 'pitching' concepts (making use if the relative motions with respect to the seabed) and 'overtopping' concepts (where the overtopping wave flow is collected in a reservoir and used to generate electricity through a turbine) [7]. Beside the fact that the concept itself might be a feasible wave energy converter, it also allows the detailed study of several types of wave energy converter systems at once. As MARIN wants to assist the development of all types of wave energy devices and not just develop the Green Water Concept, the development of knowledge is the main driver for this activity.

The Green Water Concept was initially developed using diffraction theory and Volume of Fluid (VOF) simulations. The present paper presents these simulations and the pilot model tests that were carried out to better understand the physics and the related behaviour of the system. These tests also included the initial modelling of an electrical and hydraulic Power Take Off (PTO). The accurate modelling of a PTO is an important aspect in testing of wave energy conversion concepts numerically and in a wave tank: at the moment that energy is converted into electricity in the PTO, the hydrodynamic behaviour of the structure is changing. These tests with PTO modelling will be reported in [11].

Technically the 'Green Water Concept' is a weathervaning small freeboard vessel with a water reservoir in centre, see Figure 3. The structure is moored to the seabed with a pre-tensioned wire from the seabed, over two pulleys to a pre-tension weight hanging over the stern. This single point tethered mooring also keeps the structure weathervaning head into the main wave environment. Compared to the concept presented in [5], the pre-tension weight is now hanging over the stern pulley instead of over the pulley at the bow. The advantage of this set-up is that the actual vertical motions of the weight (and resulting accelerations) are small when the structure is pitching: in pure pitch the vertical weight position is almost stationary. The concept works following 3 steps, as can be seen in Figures 3 and 4:

1. Through maximised pitch motions, the bow makes large vertical motions relative to the seabed, to which it is connected with a wire. The wire moves relative to the structure and can be attached to an electrical generator (first Power Take Off, not in Figure).
2. At the same moment, waves exceed the freeboard and green water flows onto the deck. Green water comes from the front and sides and forms a high velocity water jet. During research into the effect of green water on ship-type offshore structures [6], it was found that the flow onto the deck from the front and the sides of the bow, concentrates in a high velocity jet over the centreline of the vessel. This process can be seen in Figures 2 and 3. The resulting velocity over the centreline of the deck can be 15 to 25 m/s. This green water is able to impose high loads on structures on the deck (see the last photograph in Figure 4), but is also able to climb the slope into the reservoir of the Green Water Concept. This process will alternately occur both at the bow and stern, although the effect on the bow will be slightly stronger. This concentrated jet, together with the upward pitch motions, allows the green water to flow into a higher reservoir at the centre of the structure.
3. The green water in the reservoir then flows back into the sea through low water head turbines (second Power Take Off).

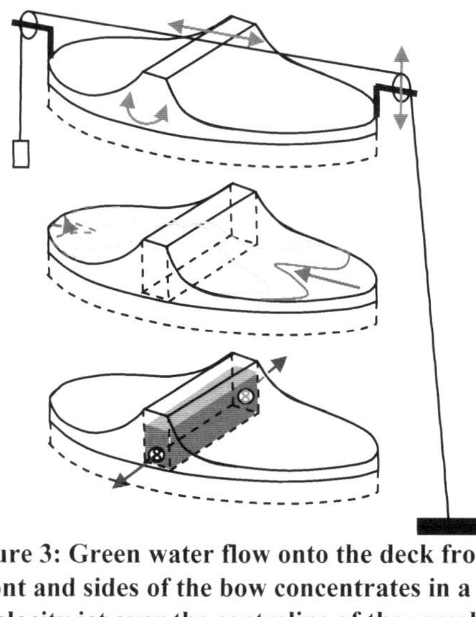

Figure 3: Green water flow onto the deck from the front and sides of the bow concentrates in a high velocity jet over the centreline of the vessel [6]

So the wave energy conversion is both a result of the motions at the bow and the green water flow out of the reservoir.

GREEN WATER CONCEPT IN PERSPECTIVE

Let us now consider the Green Water Concept in the perspective of the available power in the waves and other wave energy conversion concepts.

The main challenge of all wave energy systems is to extract as much as possible energy from the waves. For this purpose it is important to know how much energy is actually present in the waves themselves. In long-crested deep water waves the wave energy flux per unit wave crest width is P (kW/m):

$$P = \frac{\rho g^2}{64\pi} H_s^2 T$$

H_s is the significant wave height (m), T the wave period (s), ρ the density of seawater (kg/m^3) and g gravity (m/s^2). So the power of the waves is proportional to the wave period and proportional to the square of the significant wave height. As an example: a 10 second wave period with a significant wave height of 3 metres results in a 45 kW wave power potential per metre crest width.

It will be clear that, due to its high density, water is capable of transferring a lot of energy compared to wind energy systems. Consequently even small wave energy devices are capable of producing energy. By nature wave energy concepts are also low-profile compared to wind energy systems, which minimises the visual distraction if placed off-shore.

The earliest serious attempts to generate energy from the waves date from the 1970's in the time of the oil crisis. The most famous system was the 'Salter's Duck' (or officially called 'Edinburgh Duck'). However, this system was never tested offshore. New wave energy devices include the 'Pelamis', the 'Archimedes Wave Swing' and the 'Wave Dragon', of which (prototype) systems actually operate.

In an excellent summary, [2] classifies the large amount of different concepts in different groups as follows:

- *Oscillating Water Column* concepts, which are open at the bottom. The vertical motion of the water inside the structure due to the waves alternatively pressurizes and depressurizes the air inside the structure generating a reciprocating flow through a turbine at the top. Typical examples are the 'Mighty Whale' (floating) and the 'Limpet' unit on Islay (fixed).
- *Overtopping* concepts, floating or fixed to the

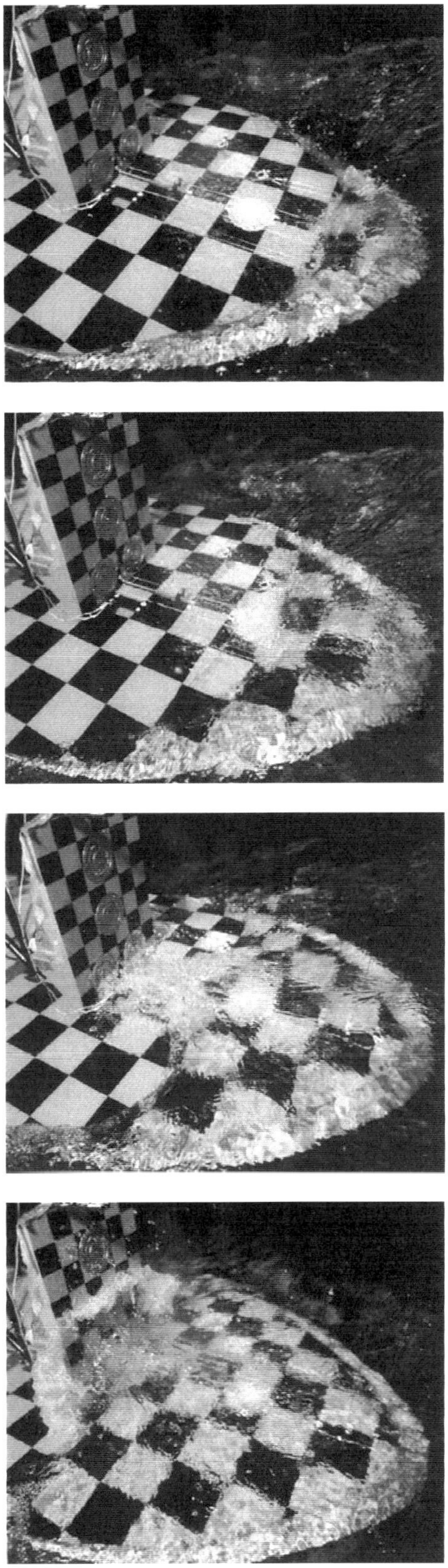

Figure 4: Green water on the deck of an FPSO [1]

shore. They collect the water of incident waves in an elevated reservoir to drive one or more low head turbines. An example on an overtopping device is for instance the 'Wave Dragon'. With a wave reflector it focusses the wave towards a ramp to fill a higher-level reservoir. Also the shore-based 'Tapchan' system is an example of an overtopping concept.
- *Heaving* concepts, which can be both floating or submerged. The heave motion is converted by mechanical and/or hydraulic systems in linear or rotational motion for driving electrical generators. An example is the 'PowerBuoy'.
- *Pitching* concepts consist of a number of floating bodies, hinged together. The relative motions between the floating bodies are used to pump high-pressure oil through hydraulic motors, which drive electrical generators. The 'Pelamis' is a typical example of this concept.
- *Surging* concepts make use of the horizontal particle velocity in a wave. This drives a deflector or generates a pumping effect of a flexible bag facing the wave front. The 'WaveRoller' is a concept in this category.

All wave energy generation or conversion concepts are facing significant challenges:
- Waves are irregular in nature. Any seastate consists of a significant range of frequencies and directions and beside that the seastates themselves change strongly in time.
- It is difficult to develop a system with optimum efficiency for a wide range of wave heights and periods.
- The concept should be able to withstand survival conditions without damage.
- The transformation of an irregular wave frequency motion (around 0.1 Hz) of the waves through electrical generators to a frequenyc typically 500 times higher is a complex process.
- The energy has to be transferred to shore, typically through subsea cables.
- Like all (renewable and normal) offshore concepts, offshore installation, maintenance and removal can be complex and costly.

We will now consider the Green Water Concept in the light of the above perspective.

The Green Water Concept has elements of overtopping, heaving and pitching concepts:
- Compared to existing 'overtopping' concepts, the overtopping is not the result of wave focussing or run up, but caused by the dynamics of the green water on a moving deck.
- 'Heaving' concepts are typically smaller buoys, where the vertical motion is only a result of the heave motion of the buoy. The Green Water Concept makes use of the combined heave and pitch motion to generate maximum vertical motions at the bow.
- Compared to the 'pitching' concepts, the Green Water Concept does not make use of hinges and related submerged hydraulics to generate energy.

Further the following strong points of the Green Water Concept can be mentioned:
- The structure is simple to build (by local shipyards).
- Easy transportation due to its ship-shaped hull. The concept is (almost) self installing after it has been towed to location. The sheaves of the electrical generator at the bow can be used to install the bottom anchor (foundation plate, suction anchor or drag anchor).
- The concept is removable for inshore maintenance and final removal (like FPSOs), which are larger challenges for fixed offshore windmills.
- Compared to smaller buoy systems, the required infrastructure will be limited less complex grid of subsea lines required) for the same amount of energy.
- The system is weathervaning, keeping the bow head into the environment
- The system can be used in deep and shallow water and is independent of the tides
- The length of the Green Water Concept will be chosen so that the period of maximum pitch excitation is close to the dominant wave period at the field location in moderate seastates. By ballasting it is possible to tune the pitch resonance period to combine the maximum wave excitation, natural period and wave period.
- In extreme weather conditions the wave periods will be longer and the short 'Inverse' concept will follow the wave slope. Further the highest waves will overtop the low freeboard structure. Though this results in extreme green water on the watertight structure, this reduces the motions, drift forces and mooring loads and protects the structure against excessive survival loads.

Based on the evaluation above, it was considered worthwile to investigate the concept further hydrodynamically. This will be done in the following phases:
1. Motion analysis and (de-)optimisation using frequency domain diffraction analysis
2. Maximisation of the green water on the deck using

an improved Volume of Fluid (iVOF) method [3-5]
3. Model testing [11]
4. Time domain motion and mooring analysis, including the evaluation of the wave energy conversion potential

The present paper focusses on Phase 1 and the initial results of Phase 2 and 3. The full Phase 2 and 3 and Phase 4 will be performed in the next stage and will be reported in later papers.

GREEN WATER CONCEPT MOTION OPTIMISATION

Methodology

The motion calculations were performed with MARIN's linear diffraction code DIFFRAC. DIFFRAC solves the linearised velocity potential problem using a three-dimensional source distribution technique. The mean wetted part of the vessel hull is approximated by a large number of panel elements. The distribution of source singularities on these panels forms the velocity potential describing the fluid flow around the vessel hull. The pressure distribution on the hull is calculated from the velocity potential. The added mass and damping coefficients, as well as the wave forces are then determined from the pressure distribution. All these calculations in DIFFRAC are carried out in the frequency domain. The added mass (a) and damping coefficients (b) and the wave load coefficients (exciting forces and moments) can be used to calculate the motion RAOs for the 6 components of the motions. As an example, the (uncoupled) equation of motion for pitch can be written as:

$$(I_{yy} + a_\theta)\ddot{\theta} + b_\theta\dot{\theta} + c_\theta\theta = M_\theta$$

As mentioned before, the length of the Green Water Concept will be chosen so that the period of maximum pitch excitation is close to the dominant wave periods at the field location in moderate seastates. As a starting point the length of the Green Water Concept was chosen at 60m. For normal ship-type offshore structures such as FPSOs, the pitch resonance period is typically much shorter than the period of maximum wave excitation, resulting in limited pitch motions. However, for the Green Water Concept the mass moment of inertia can be chosen such that the natural pitch period shifts to the period of maximum wave excitation, resulting in more resonant-type motion. At resonance the inertia and spring terms in the equation of motion are cancelling eachother, resulting in a response dominated by wave excitation and system damping (which includes energy generation).

Variations

To maximise the pitch motions (local vertical motions at the bow) and relative wave motions (Figure 5, as input to the green water), a range of structural shapes and dimensions was investiaged.

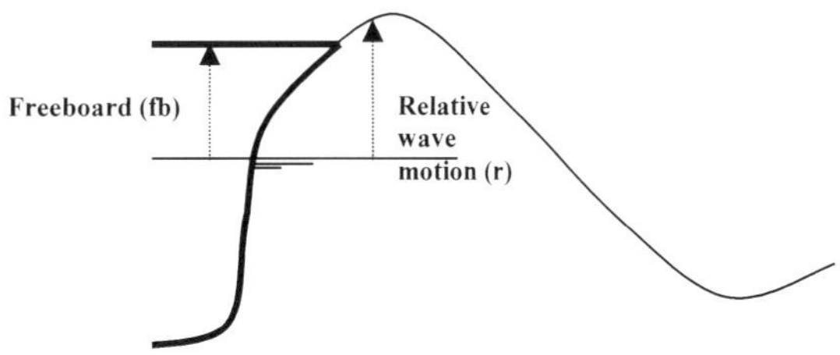

Figure 5. Definition of relative wave motion (r) and freeboard with respect to the waterline in calm water (fb)

The first group (Figure 6) consisted of 3 different ellipsoids with different length over beam (L/B) ratios of 3, 5 and 7:

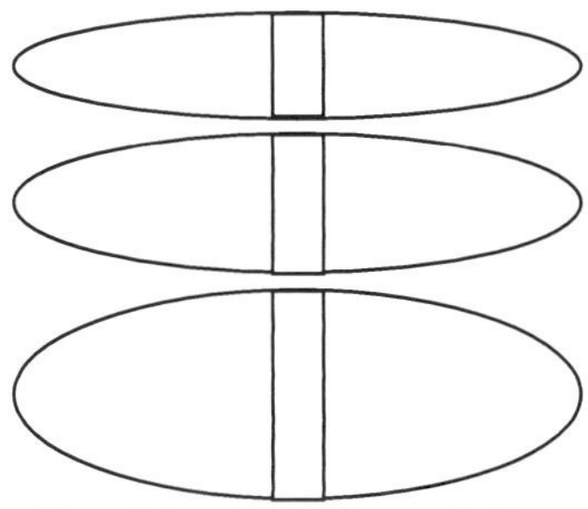

Figure 6: Ellipsoids with different L/B ratios of 3, 5 and 7

The second group (Figure 7) included more triangular structures, as it was concluded in [1] that triangular structures can have (un)favourable large motion characteristics and that the triangular shape of the deck can result in a really focussed green water flow on the deck.

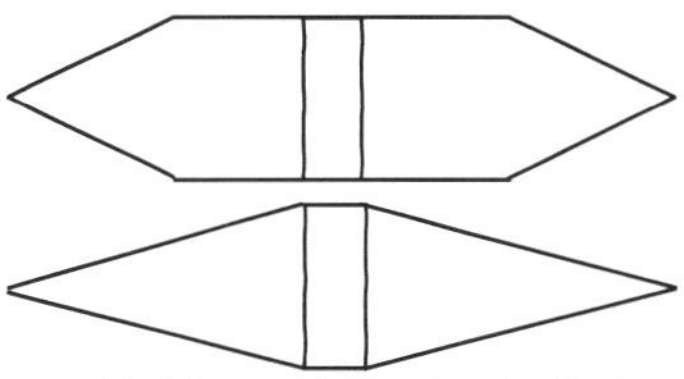

Figure 7: More triangular hull shapes

All structures had simple vertical sides over the full depth of the structure. In Figures 8, 9 and 10 an overview is given of the element distributions used for the diffraction analysis.

Another variation often investigated in naval architecture is the ratio between the beam and draft (B/T) of the structure. With increasing draft the wave damping decreases, whereas the wave excitating is not varying that much. This typically results in higher heave and pitch motions with larger drafts of ship-type structure (with extremely large drafts, as used for Spars, the wave excitation reduces significantly, almost eliminating the heave motions). For all the ellipsoids, three B/T-ratios are used of 3 (highest draft), 5 and 7 (smallest draft). In a later stage one ellipsoid was added with a B/T ratio of 1 (Body 0). For one triangular structure also a B/T variation was applied.

For the radius of pitch gyration (kyy) a value of 15m (25% of L) was chosen as base case. For Body 1 (Ellipsoid, L/B=3, B/T=3) and Body 12 (Triangular, L/B=3, B/T=3) a realistic variation of radii of gyration was applied (using ballast) of kyy=10, 12, 15, 18 and 20. Table 1 gives an overview of the resulting main dimensions of all variations with its reference colours and symbols that will be in the presentation of the results.

			B/T	L/B	L	B	T	D	Displ	GMl	GMt
			[-]	[-]	[m]	[m]	[m]	[m]	[tonnes]	[m]	[m]
Body 0	•	Ellipsoid	1.0	3.0	60.0	20.0	20.0	21.0	19321.0	11.2	1.2
Body 1	▪	Ellipsoid	3.0	3.0	60.0	20.0	6.7	7.7	6440.3	33.6	3.7
Body 2	◊	Ellipsoid	5.0	3.0	60.0	20.0	4.0	5.0	3864.2	56.1	6.2
[illegible]	[illegible]	[illegible]	[illegible]	[illegible]	[illegible]	[illegible]	[illegible]	[illegible]	[illegible]	[illegible]	[illegible]
Body 4	x	Ellipsoid	3.0	5.0	60.0	12.0	4.0	5.0	2318.5	56.0	2.2
Body 5		Ellipsoid	5.0	5.0	60.0	12.0	2.4	3.4	1391.1	93.4	3.7
Body 6		Ellipsoid	7.0	5.0	60.0	12.0	1.7	2.7	993.6	130.7	5.2
Body 7	+	Ellipsoid	3.0	7.0	60.0	8.6	2.9	3.9	1182.9	78.5	1.6
Body 8		Ellipsoid	5.0	7.0	60.0	8.6	1.7	2.7	709.7	130.8	2.6
Body 9		Ellipsoid	7.0	7.0	60.0	8.6	1.2	2.2	507.0	183.1	3.7
Body 10	□	Triangle	3.0	5.0	60.0	12.0	4.0	5.0	2214.0	46.8	2.5
Body 12	*	Triangle	3.0	3.0	60.0	20.0	6.7	7.7	4375.5	22.6	2.8

Table 1: Main particulars of all variations

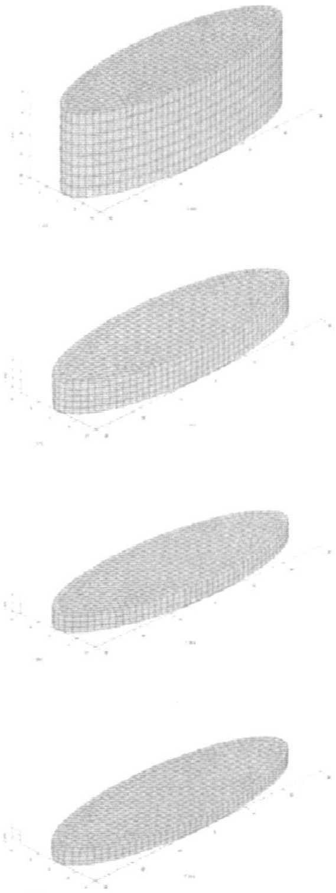

Figure 8: Element distributions with an L/B of 3 and B/T of 1, 3, 5 and 7 (from top to bottom: Body 0, 1, 2 and 3)

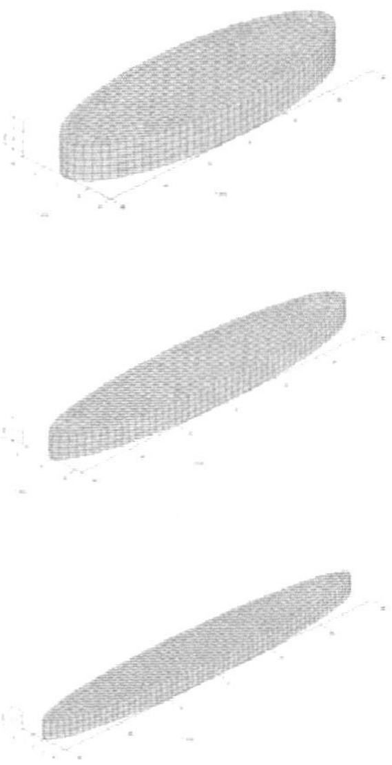

Figure 9: Element distributions with an L/B of 3, 5 and 7 and B/T of 3 (from top to bottom: Body 1, 4 and 7)

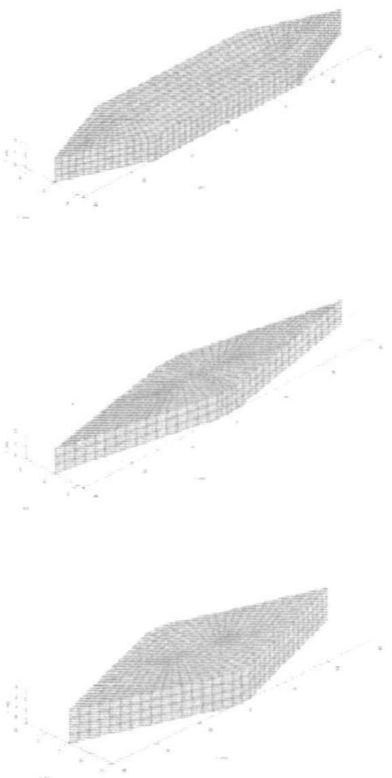

Figure 10: Element distributions with the triangular structures (from top to bottom: Body 10, 11 and 12)

Results

In the next few pages an overview will be given of the results of the systematic series of calculations carried out. By consistent colour codes and symbols it is tried to keep the overview as good as possible. The results of some bodies (5, 6, 8 and 9) are not presented to limit the size of the paper and because the results of the other bodies cover the trends found for these bodies.

As a starting point, Figure 11 presents the pitch motion RAOs for different groups of bodies. The upper graph presents the results of the B/T variation for bodies 1, 2 and 3 with the same width. It will be clear that the body with the largest draft (body 1) shows the largest pitch response (peak at almost 9 degrees per metre wave amplitude). This is a result of the lowest pitch (wave) damping combined with an almost similar wave exciting moment as for the other drafts. This will be discussed later in more detail. The middle graphs shows the sensitivity for the L/B ratio (with constant B/T ratio) for bodies 1, 4 and 7. It will be clear from this graph that the widest structure (body 1) has the largest pitch response. For triangular bodies 11 and 12 the pitch response (bottom graph) is even more extreme. For body 12 a peak response of 19 degrees/m is observed, clearly a result of resonant behaviour.

Figure 12 shows a similar comparison for the heave response. These graphs show that only body 1 and 12 really show resonant heave response around 0.8 rad/s.

Figure 14 combines the heave motions for bodies 1, 2, 3 and 12 with the pitch motions, to determine the absolute vertical motion at the anchor line fairlead at the bow of the structure (assumed at +35m, 5 m in front of the bow). This Figure confirms the large motions of bodies 1 and 12. In a wave of 1.0m, the fairlead point of body 1 moves more than 5.0m and body 12 even more than 10.0m. This is without the effect of a generator extracting energy for the system and reducing the motions accordingly. Also non-linear effects (expected with these large motions) are neglected in this calculations.

Figure 13 then presents the relative wave motion RAOs at the bow of the structure. Some interesting trends can be observed:

- For bodies 1, 2 and 3 the same type of trend is observed as for the pitch motions. This can be well understood because of the strong relation between (out of phase) pitch motion and relative wave motion.
- In long waves the structures follow the wave slope (no relative wave motions), in short waves the structure is not moving anymore and the relative wave motion is purely a result of the incoming wave (RAO value of 1.0) and the reflected wave (RAO value of +1.0 for full bows and +0.0 for thin triangular bows). This trend is clear in the results. The high frequency limit for bodies 1, 2 and 3 (top graph) is close to 2.0 (almost standing waves), for the less wide bodies 4 and 7 this reduces (middle graph) and for the thin triangular bows 10, 11 and 12 it is close to one.
- The relative wave motions for bodies 1 and 12 are remarkable. With an input wave of 1.0m, the relative motion of body 1 is 5.5m and for body 12 this is even 11m. It should be noted that this neglects the possible effects of the green water on the motions [1].

It is important to note that the motion response of the structure as observed, is a resonant response to the wave excitation. At resonance, the spring and inertia terms in the equation of motion cancel each other, which implies that the response is determined by the balance between the wave excitation and damping only (it should be noted that in the present calculations the damping effects of energy generators is neglected, so that the actual motion response will be lower).

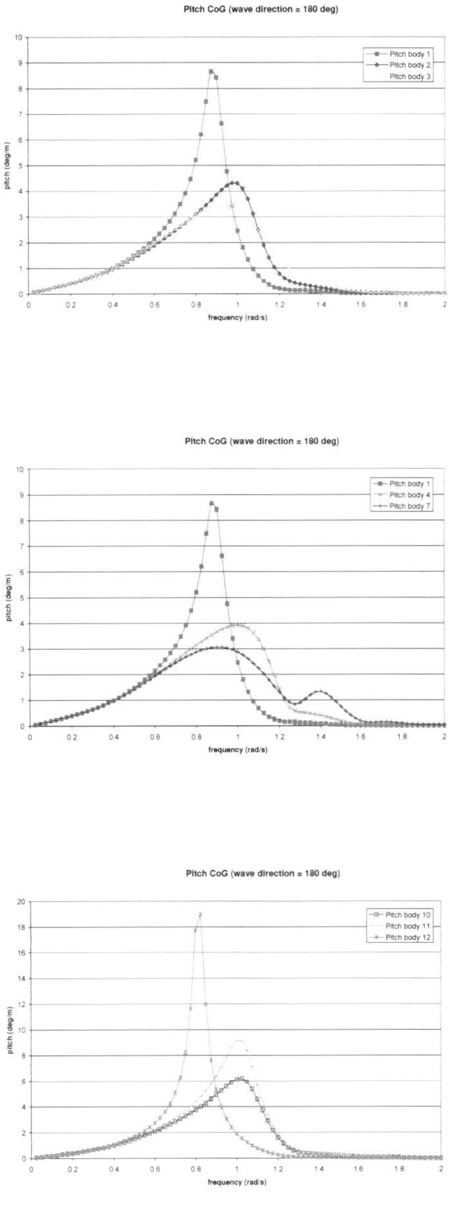

Figure 11: Pitch motion RAOs for different groups of bodies

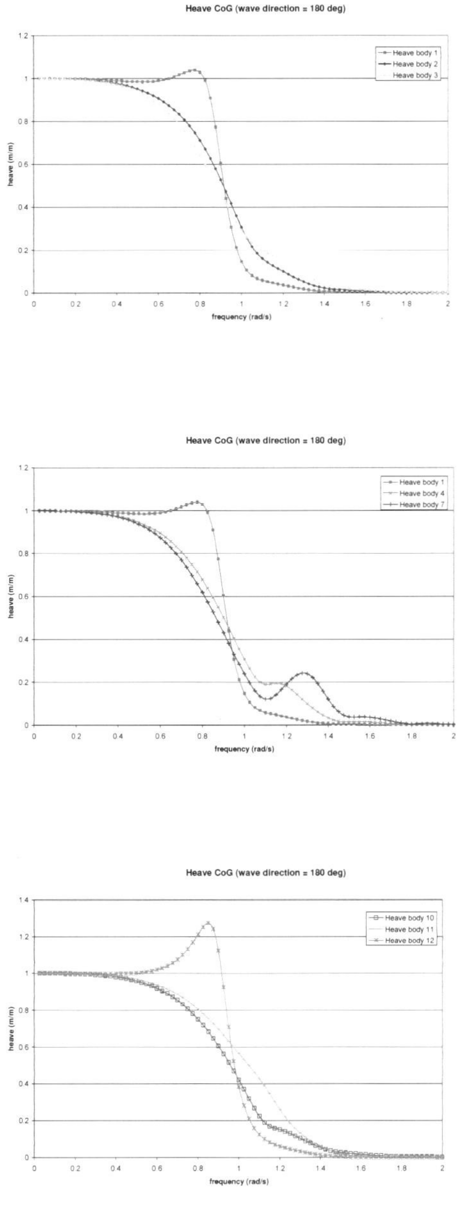

Figure 12: Heave motion RAOs for different groups of bodies

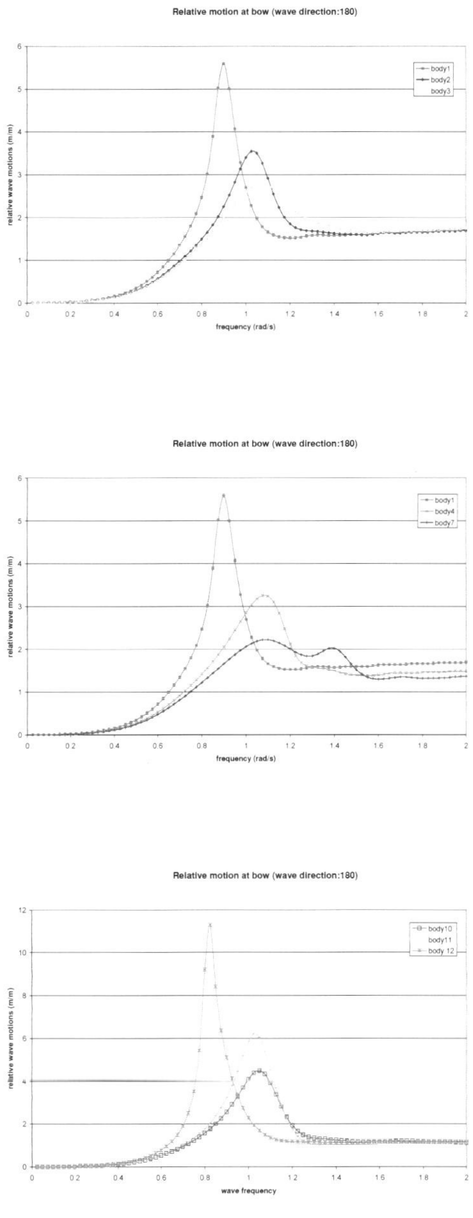

Figure 13: Relative wave motion RAOs for different groups of bodies

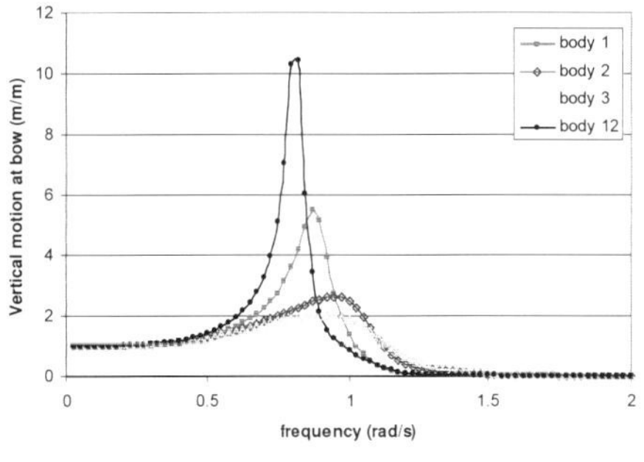

Figure 14: Vertical motion at the bow for bodies 1, 2, 3 and 12

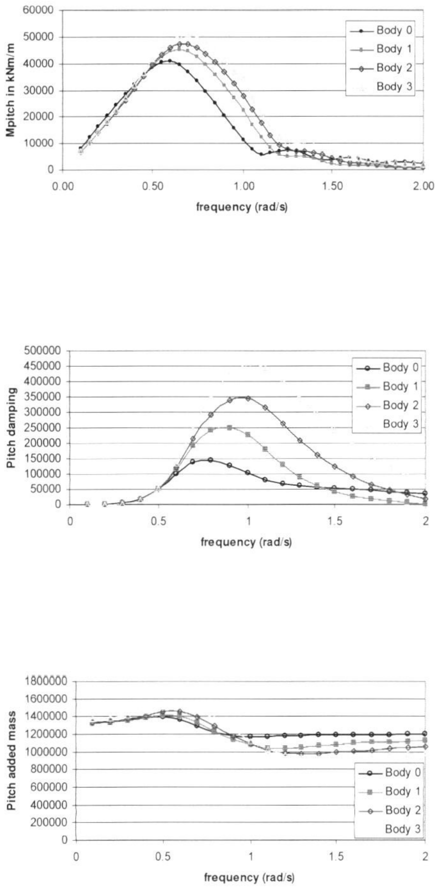

Figure 15: Wave exciting moment, pitch damping and pitch added mass for body 0, 1, 2 and 3

Therefore, it is now time to investigate the background of the observed differences in more detail. In this we will focus on elliptical bodies 1-3 and triangular body 12. Figure 15 shows the pitch exciting moment, pitch damping and pitch added mass (from top to bottom) for body 1 (B/T=3), 2 (B/T=5) and 3 (B/T=7.0). In a later stage a body 0 was added with an even large draft (B/T=1). The following typical trends can be observed:

- The pitch exciting moment (top graph) reduces only slightly with increasing draft. Only from B/T=3 (body 1) to B/T=1 (body 0) there is a signficant reduction in wave excitation. The reduction of the wave excitation is explained by the reduced orbital motions at the level of the bottom of the structure (exponential decay).
- The reduction of the wave damping (middle graph) is, however, much stronger with increasing drafts. This is a result of the fact that the wave damping (radiation damping) strongly reduces when the distance of the bottom of the structure to the free surface increases.
- For the structure with the largest draft, the high frequency limit of the damping becomes larger again. This is a result of the contribution of the vertical sides of the structure to the wave damping, which for a structure with a larger draft becomes significant.
- The added mass becomes more and more constant with increasing draft, as the free surface effects in the added mass become less important.

So the increased response at increasing draft can be explained by the fact that the wave excitation reduces less with the increasing draft than the wave damping. A reduced wave damping is attractive as this implicates that less energy is dissipated by wave radiation and can be used for actual wave energy conversion. For the moment body 1 is used for further work (as the draft of body 0 is considered to be a bit extreme, also considering the costs of steel).

Figures 16 and 17 now present the wave exciting pitch moment and pitch damping for bodies 1, 4 and 7 (above) and bodies 10, 11 and 12 (below). Compared to body 1, the less wide bodies 4 and 7 have a smaller wave exciting moment as expected. The wave damping also reduces, but at a lesser rate.

Considering the large motion response of body 12 in Figure 14, at first sight it is surprising that the wave exciting moment on this body is so low compared to (for instance) body 1. This lower excitation can be understood from the triangular shape of the bow. The larger response can be explained by the even smaller pitch damping as can be observed in Figure 17.

So it is concluded that the actual motion response is a result of the balance between the wave exciting moment and (wave) damping. However, this ratio is dependent on the frequency as well. Further, the actual motion response is determined by the resonant period, in which the spring (GM) and inertia terms play important roles. To understand these relations one step better, Figure 18 shows the pitch motions, pitch wave excitation (/10000) and pitch damping (/30000) for bodies 1, 2, 3 and 12 (from top to bottom). For sake of comparison, the same vertical scale has been used. The following can be observed if we study these graphs carefully:

- The pitch excitation peaks at a wave frequency of 0.7 rad/s (T=8.98s), which represents a deep water wave length of approximately 125m. This is surprisingly long considering the length of the structures of 60m. The length of the Green Water Concept can be adjusted to the range of existing wave periods at a certain location. The present 60m of length can be considered as an upper limit based on these results.
- The reducing response with decreasing draft of bodies 1, 2 and 3 is clear from the graphs. However, this is not just an effect of the increasing wave damping, but also an issue of natural periods. Table 2 shows the longitudinal and transverse GM values and natural pitch periods for these bodies (with kyy=15m for all). The natural periods can be determined from the following expressions:

$$T_{natural} = 2\pi\sqrt{\frac{I_{yy} + a_{\theta}}{\rho g \nabla GM}}$$

$$I_{yy} = k_{yy}^2 \rho \nabla$$

It will be clear from this table that the lighter bodies 2 and 3 have smaller natural periods. As a result, their maximum pitch response occurs further away from the point of maximum pitch exciting moment (and minimum pitch damping).

Body	GMl	GMt	Tnatural	ωnatural
	[m]	[m]	[s]	[rad/s]
1	33.65	3.72	6.84	0.92
2	56.08	6.21	6.03	1.04
3	78.51	8.69	5.69	1.10

Table 2: Longitudinal and transverse GM values and natural pitch periods and frequencies for bodies 1, 2 and 3.

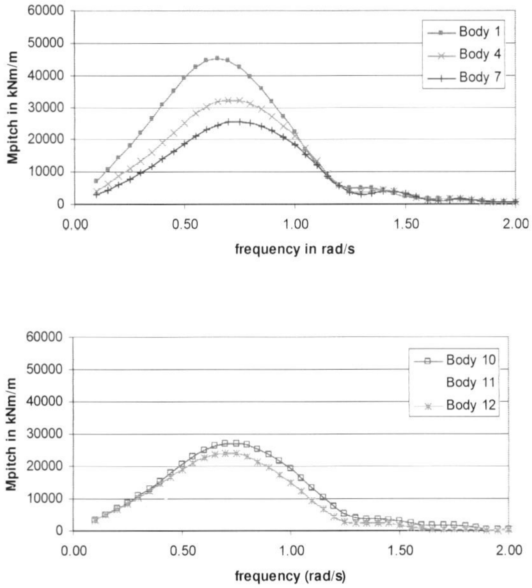

Figure 16: Wave exciting moment for bodies 1, 4 and 7 (above) and bodies 10, 11 and 12 (below)

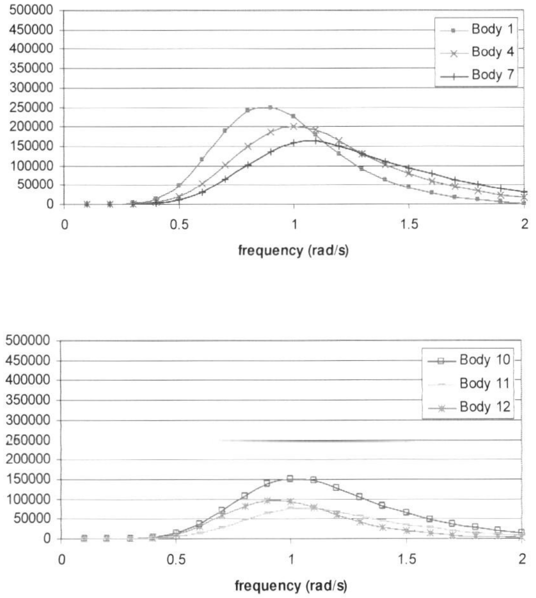

Figure 17: Pitch damping bodies 1, 4 and 7 (above) and bodies 10, 11 and 12 (below)

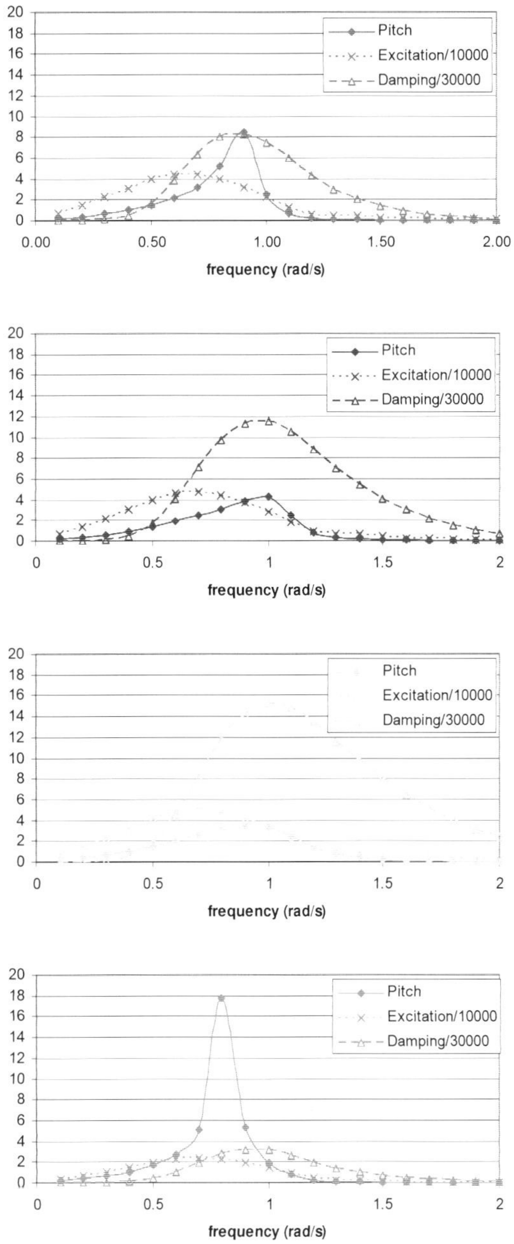

Figure 18 Pitch motions, pitch excitation (/10000) and pitch damping (/30000) for bodies 1, 2, 3 and 12 (from top to bottom)

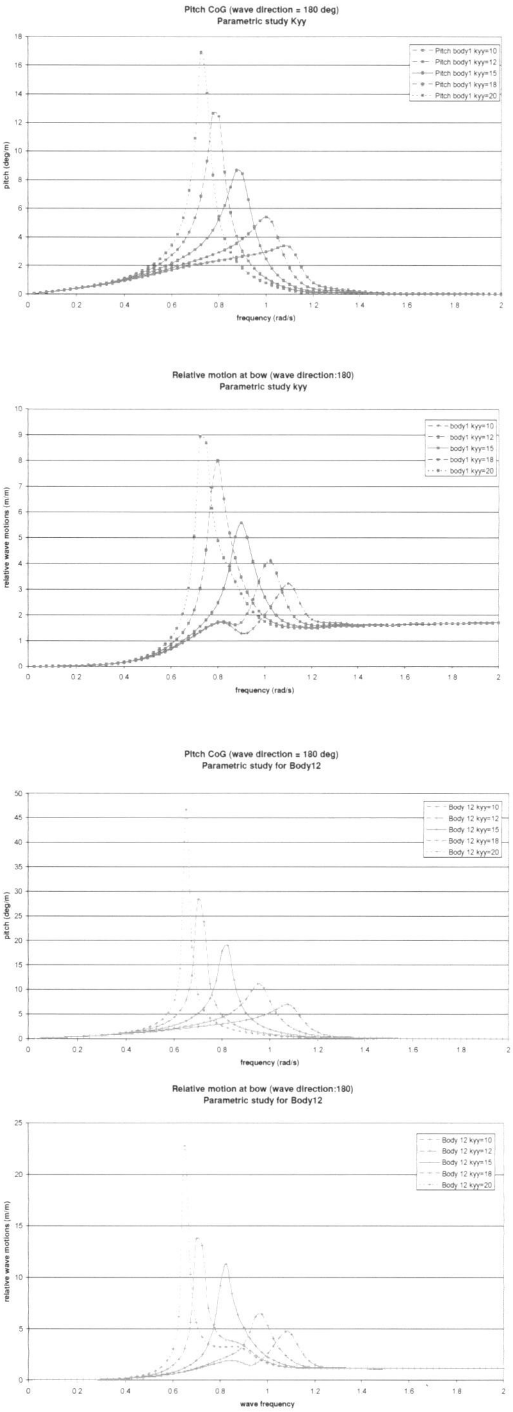

Figure 19 Effect of kyy variation on pitch motions and relative wave motions of body 1 (above) and body 12 (below)

So apart from the importance of a maximum pitch exciting moment and minimum wave damping, the natural period of the response is very important. As can be seen for body 1 (top graph of Figure 18) for this body the pitch peak response is not the same as the point of maximum pitch excitation. For body 12 this is better, but still not exact.

Therefore, a systematic variation of the pitch radius of gyration was carried out for bodies 1 and 12. Figure 19 shows the effect of this kyy variation on the pitch motions and relative wave motions of body 1 (above) and body 12 (below). A realistic range of kyy was chosen (based on possible water ballast and structural weight): 10m, 12m, 15m, 18m and 20m. This resulted in natural pitch periods of 5.64s-8.23s (1.11-0.76 rad/s) for body 1 and 5.73s-9.29s (1.1-0.68 rad/s) for body 12. The results in Figure 19 make clear that this allows the tuning of the peak response to the area of maximum wave excitation (and minimum wave damping) and to adjust the 'Inverse concept' to the period of the incoming waves. Variation of the pitch natural period with the GM value (height of the centre of gravity) is not a real option, as the transverse GM values are limited (see Table 2) and needed for transverse stability.

So far the motion behaviour has only been studied based on (RAO) graphs. To better understand the

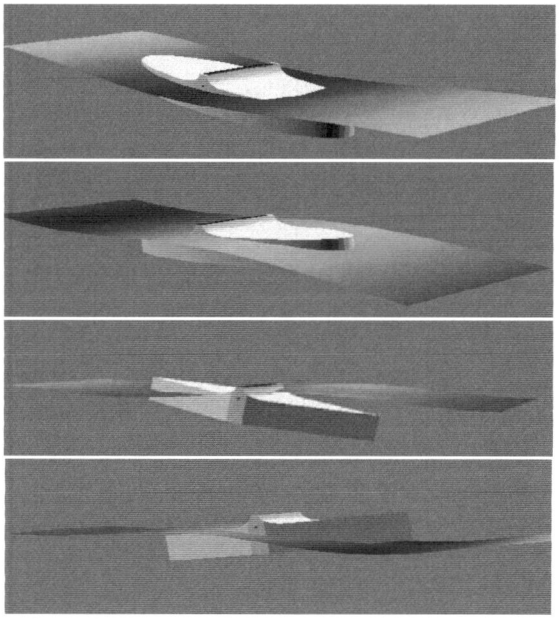

Figure 20 Visualisation of the behaviour of body 1 (top) and 12 (bottom) using the motions and disturbed wave field of the linear diffraction analysis (above calm level body part only present in visualisation)

implications of these results, it is also useful to visualise the waves and motions. This is done in Figure 20 for body 1 and 12 using the motions and wave field of the linear diffraction analysis.

The Figures confirm the large motions and relative wave motions presented in previous graphs. The bow makes large vertical motions and both the bow and stern submerge as a result of the large relative wave motions.

It is important to note that in these Figures everything above the calm level is only added for visualisation purposes and neglected in the actual underlying diffraction analysis. Especially the water shown on top of the deck is unrealistic as a result. These Figures show that green water will flow onto the deck, but not how. For this purpose the next series of simulations with the improved Volume of Fluid (iVOF) method are necessary.

INITIAL IVOF SIMULATION OF GREEN WATER

The iVOF method in the ComFLOW program (developed by the University of Groningen, RuG) is based on the Navier-Stokes equations for an incompressible, viscous fluid. The equations are discretised using the finite volume method. The displacement of the free surface is done using the Volume of Fluid method first introduced by Hirt and Nichols [3-5]. To avoid small droplets disconnecting from the free surface, the iVOF-method is combined with a local height function. The iVOF method has already been used for a number of applications, like sloshing on board tumbling spacecraft, and blood flow through arteries. Maritime applications are sloshing in anti-roll tanks, green water flow on the deck, TLP response to extreme waves and falling objects in calm water [1, 3-5].

In [1,3] the results were shown of the simulation of the flow of green water over the deck of an FPSO and the resulting impact on deck structures. The computational domain was limited to the area on and around the deck. The freeboard exceedance around the deck was used as boundary condition and the deck was not moving in this approach. For the present initial simulations the same quasi-static approach has been used. In the next phase also the motions of the structure will be taken into account, as was done for an FPSO in [4] and TLP in [5]. However, the good results presented in Figures 21 and 22 (from [1]) for the flow onto the deck and the impact loading on the structure on the deck (see also Figure 3), justify the use of this approach at this stage of the Green Water Concept development.

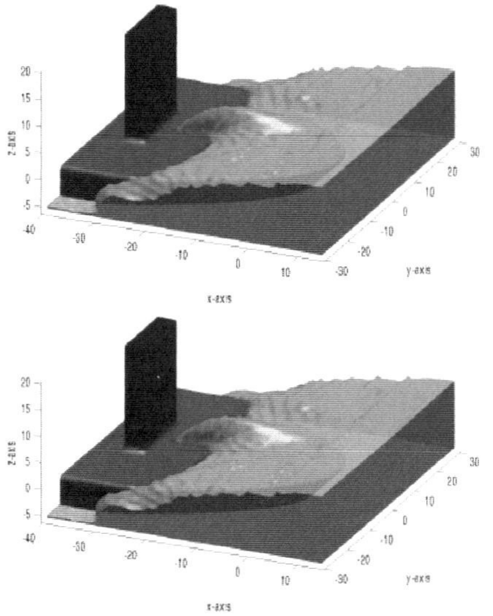

Figure 21: Visualisation of a green water simulation on an FPSO with the iVOF method [1], compare to the tests in Figure 3.

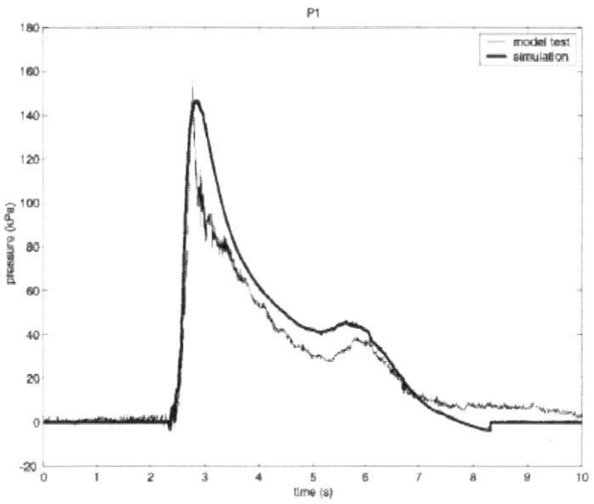

Figure 22: Comparison between the simulated and measured green water impact on the structure on the deck [1].

Figure 23 shows the description of bodies 1 and 12 as input for the iVOF simulation. The main body, the slopes, the reservoir and the outflow openings are modelled.

For the present initial simulations only half the body was used. As starting/boundary condition of the simulation a stationary 'wall' of water around the structure is chosen of 5.0m high above the deck level. Considering the freeboard level assumed (2.0m), this requires a relative wave motion of 7.0m. Based on the relative wave motion results presented in Figure 19, this will regularly occur in significant wave heights in the order of 2-3m.

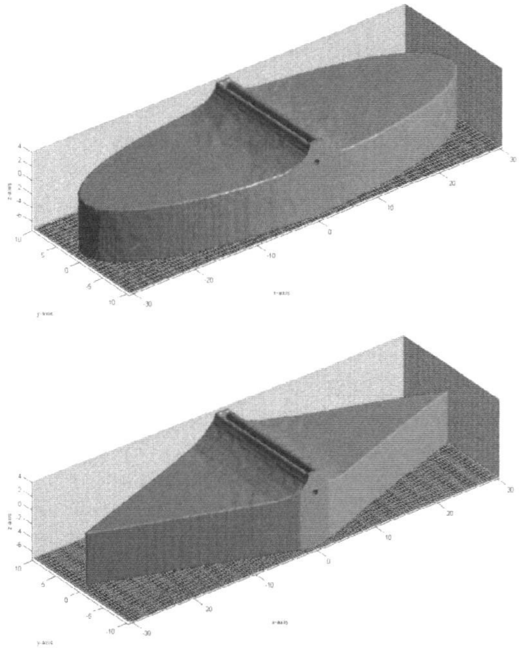

Figure 23: Modelling of body 1 and body 12 for the iVOF simulation.

The different stages of the flow onto the deck are shown in Figure 24 for the elliptical body 1 and in Figure 25 for the triangular body 12. The Figures start with the initial condition at the beginning of the simulation. Then both bodies show the required behaviour where water from the fronts and sides of the bow flows towards to the centerline, where it forms a high velocity water jet over the deck. The water runs up the slope of the deck and then flows into the top of the reservoir. The reservoir collects the water and the water flows out of the openings at the side, at which lower head turbines will be placed. At present a simple squared reservoir was used. In reality an optimised vertical funnel shaped structure will be used to maximise the pressure head.

It will be clear from Figures 24 and 25 that the shape of the deck in plan view has a large effect on the flow on the deck itself. The triangular shape of body 12 with its sharp apex seems particularly effective to combine the two waterfronts of the sides into a jet along the centreline.

It should be noted that in reality the green water flows onto a moving deck. The pitching motion of the structure (bow up after the green water has flown onto the deck) will increase the flow velocity of the water over the deck into the reservoir. Further the green water flow will occur both from the bow and the stern. As the green water flow onto the deck is mainly a dam-breaking type of flow [1], the green water velocities are mainly a result of the local relative wave elevation and less influenced by the orbital motion of the waves. In that respect similar flow from the bow and stern is expected, except for the difference in relative wave motions (which are affected by wave reflection, although this will be small for the thin body 12).

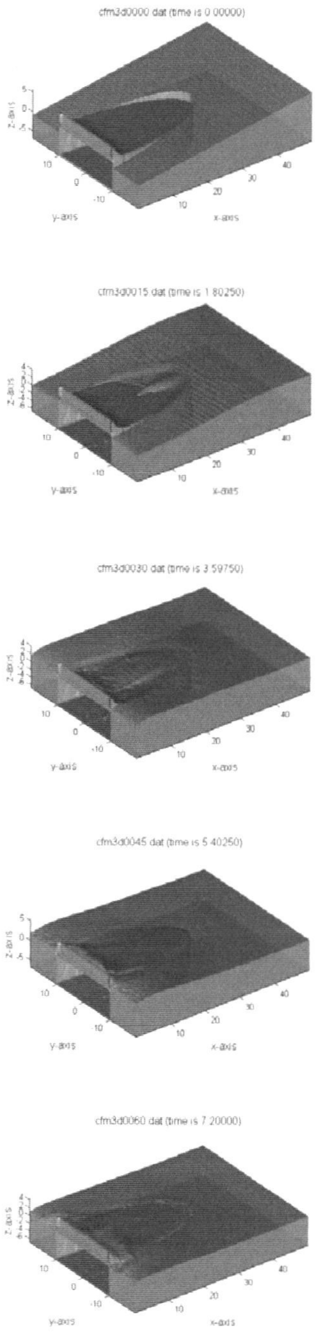

Figure 24: Flow onto the deck and into the reservoir for body 1 with a freeboard exceedance of 5.0m

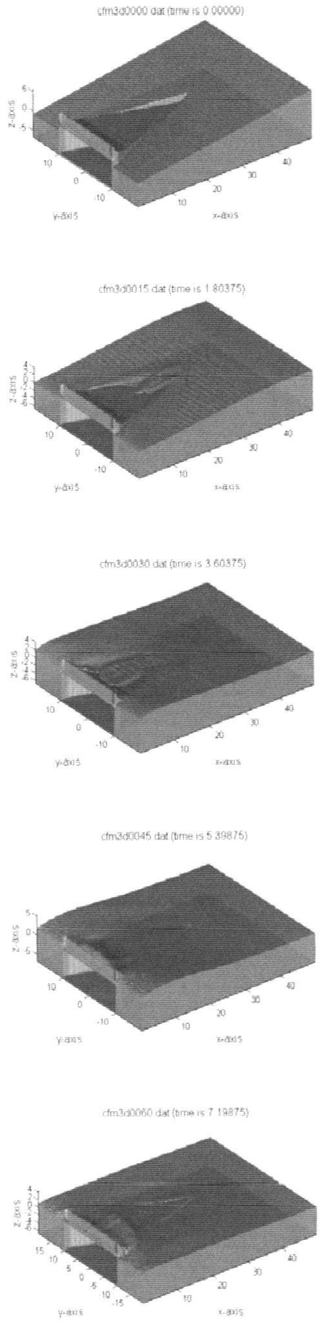

Figure 25: Flow onto the deck and into the reservoir for body 12 with a freeboard exceedance of 5.0m

MODEL, INSTRUMENTATION AND SETUP

Pilot model tests are performed to validate these calculations in a situation that is clearly non-linear (with water flowing onto the deck in high waves), to understand the physics and to determine the effect of the Power Take Off (PTO) on the motions.

For the tests a careful 3D design was made of the model, as can be seen in Figure 26. To have a stiff and watertight model, PVC was used as material. The PVC was welded together and rubber rings were used for removable parts such as decks. Steel and lead ballast weights were used to achieve the correct weight distribution as given in Table 3. The scale was 1:30 and the kyy=16.45m.

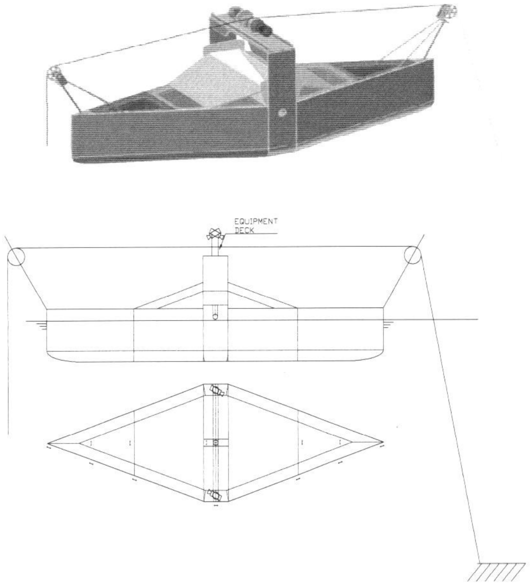

Figure 26: Design of the Green Water Concept model

Designation	Symbol	Unit	Magnitude
Length between perp.	L_{pp}	m	60.0
Beam	B	m	20.0
Depth	D	m	9.0
Draught (even keel)	T_{mean}	m	7.0
Displacement	Δ	tonnes	4628
CoG above base	KG	m	5.21
CoG from station 10	LCG	m	0.0
Transverse GM	GM_T	m	1.11
Transverse radius in air	k_{xx}	m	5.51
Longitudinal radius in air	k_{yy}	m	16.45
Horizontal radius in air	k_{zz}	m	16.45
Pitch period	T_θ	s	8.4
Water depth	WD	m	30

Table 3 Main particulars

Beside this, the following was modelled:

- A streamlined pre-tension weight hanging on the wire at the stern side, consisting of a conical top and bottom with rings in between. The weight could be changed by removing or adding these metal rings (see Figure 29). For the tests 30t to 50t was used.
- Two slopes/ramps were prepared to fill the reservoir: a high ramp up to a level of 4.5m above the deck and a low ramp up to a level of 3.0m.
- The water reservoir had a capacity of approximately 500m^3 (full scale). It was made with downward slopes to the centre to keep the pressure at the bottom high, also with low filling rates (needed for the turbine). The turbine itself was not tested so the tests had either a free outflow or no outflow. The tubes below the deck can be seen on the construction pictures in Figure 27. The outflow rate was approximately 9m^3/s (full scale) with a fully filled reservoir.

Figure 27 Reservoir with outflow on the sides trough a pipe below deck

- Two pulleys at the bow and stern, with force measurement transducers between the pulley and the structure (Fx, Fz).
- To have the Power Take Off (PTO) accessible during the tests, it was placed above the reservoir (in a real situation it would be placed below deck together with the wires). In the next section the PTO modelling will be discussed in detail. At this moment it is important to note that the PTO is a rotating engine/generator with two drums. To prevent a slipping wire on the drum, the wire from the bow side was connected on top side of one drum, whereas the stern wire to the pre-tension weight was connected to the bottom side of the other drum. Two vessel fixed cameras were used on the model for green water flow visualisation on the deck
- The relative wave motions were measured around the structure and the green water height on the deck in front of the ramp, see Figure 29.
- The mooring restoring force (F_x) is a result of the horizontal component of the pre-tension in the mooring wire ($F_{pretension}$). With a horizontal displacement, this horizontal component becomes larger. It can easily be derived that the horizontal stiffness depends on the pre-tension level, water depth (WD) and horizontal displacement according to:

$$F_x = F_{pretension} \cdot \frac{x}{\sqrt{WD^2 + x^2}}$$

In Figure 28 the resulting force-displacement curve is given for the 30m water depth with a 300kN pretension.

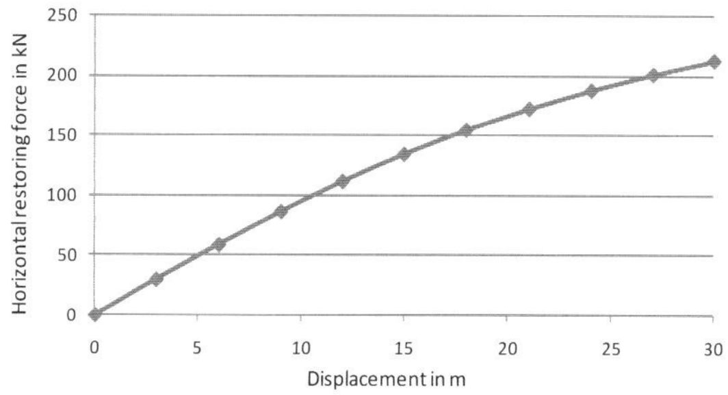

Figure 28 Mooring stiffness with a 300kN pretension in 30m water depth

The final model outside and inside the basin is presented in Figures 29 and 30. Heavy weights at the basin floor were used as bottom anchor.

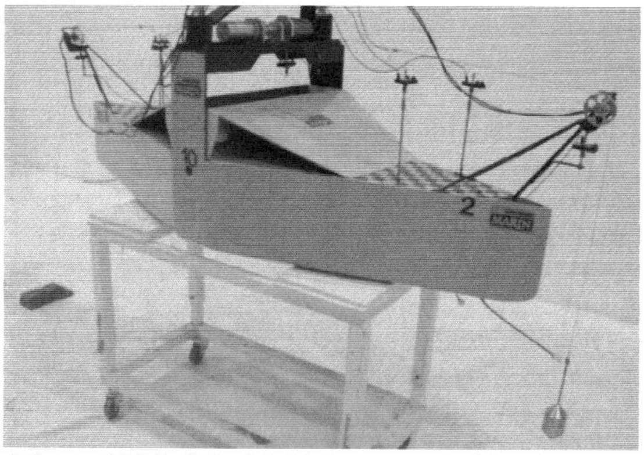

Figure 29 Model with high ramp outside the basin

Figure 30 Model with low ramp inside the basin

Careful study of the differences between Figure 29 and 30 shows that the model in the basin has an additional guiding wheel in front of the bow (below the pulley at deck level). This was needed to keep the wire from running off the pulley when transverse motions occurred during the tests.

In [11] the results using the Power Take Off will be presented. Function of a Power Take Off (PTO) is to convert kinetic or potential mechanical energy from a structure into electric energy.

During this conversion forces are exerted by the PTO on the structure. These forces influence the structure's hydrodynamic behaviour. That is why it is important to model the PTO behaviour in numerical and physical testing of wave energy converters. As an initial step in better modelling of PTOs during wave tank tests, the pilot tests on the 'Green Water Concept' also included the modelling of a PTO.

OBSERVATIONS

As indicated in the previous section, tests were performed in a series of regular and irregular waves. First tests were carried out without the PTO modelling to better understand the overall hydrodynamic behaviour. These will be presented below. In the next section the effects of the PTO will be described.

The tests confirmed the high pitch motions and relative wave motions in low waves as predicted in [5]. Also the concentrated high velocity water green water jet along the length of the structure (needed to fill the reservoir in the center) as simulated in Figure 25 was clearly observed.

Figure 31 shows the green water flow onto the deck, this concentrated high velocity jet over the centreline and flow into the reservoir in a regular wave of 3.0m high. Significant amounts of green water reached the reservoir in each oscillation, but it was difficult to quantify these amounts as there was a continuous high outflow at the same moment.

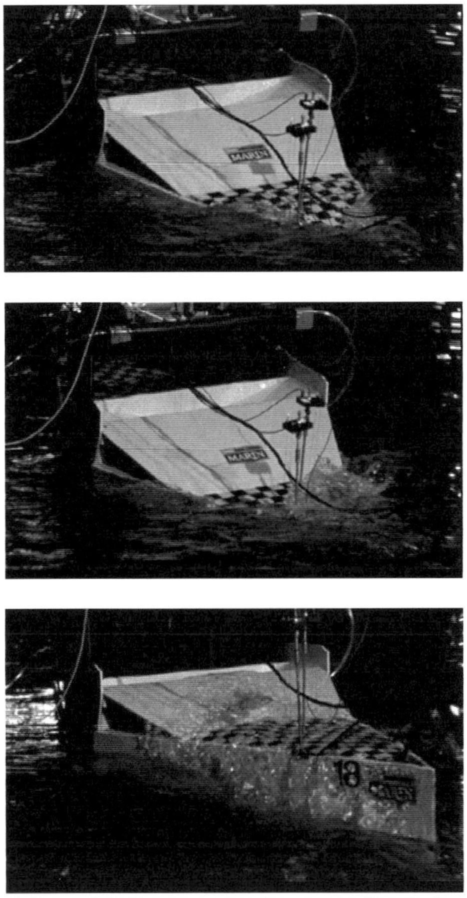

Figure 31 Green water flow onto the deck, concentrated high velocity jet over the centreline and flow into the reservoir

The testing also showed an unexpected behaviour: as sketched in Figure 32, the green water that just did not reach the reservoir resulted in a returning flow on the ramp that impacted on the incoming green water flow of the next wave cycle. This affected the effective flow of green water onto the ramp significantly. This was not identified during the simulations as these considered only one flow event on a fixed structure. This showed the value of the model testing for the evaluation of a concept.

A possible solution for this problem is an additional ramp level and triangular flow guides below it to guide the returning flow to the sides. This is shown schematically in 33 and will be investigated in future model testing campaigns.

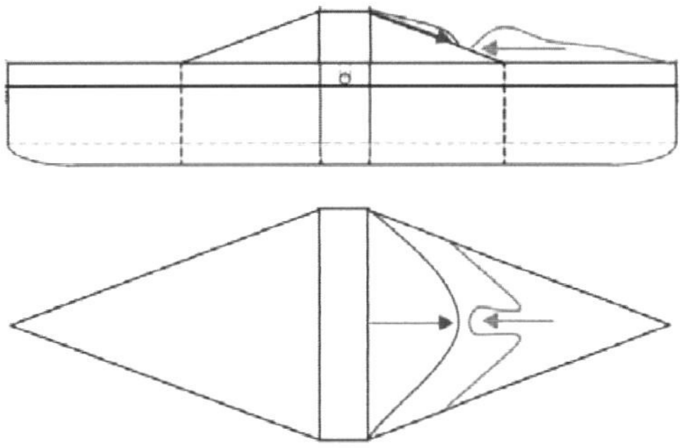

Figure 32 Meeting of incoming flow and returning flow (that just did not reach the reservoir), reducing the effective flow onto the ramp

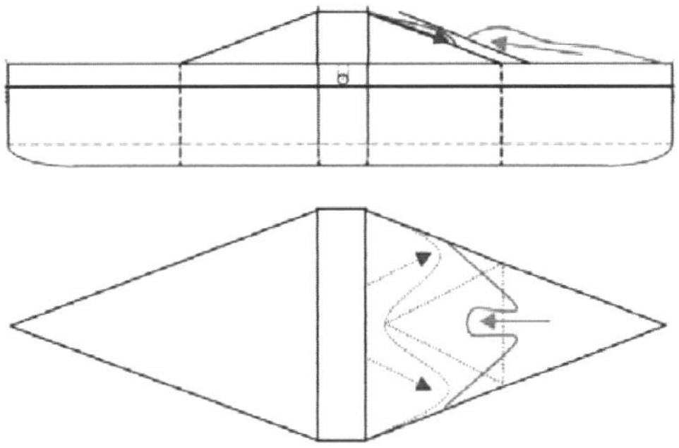

Figure 33 Possible solution with an additional ramp level and triangular flow guides below it to guide the returning flow to the sides

We will now look to the test results in a more quantitative way. Figure 34 shows the time traces in a regular wave of H=3.0m without PTO modelling. The following signals are shown:
PITCH_MOT: Pitch motion
Z_REF: Vertical motion of the forward pulley
REL_FORE1: Relative wave motion at the bow
REL_RAMP_F: Green water height on the fore ramp
REL_RAMP_A: Green water height on the aft ramp

The following can be observed:

- With a wave height of only 3.0m (~1.5m single amplitude), the single amplitude pitch motion is already 7.6 degrees and the vertical motion at the bow 6.0m.
- There is already approximately 4.5m green water height on centre of the deck in these waves (the freeboard is 2.0m).
- The relative wave motion measurement at the bow is truncated as the wave probe is completely submerged.
- At the stern the relative wave motions and amount of green water on the deck is much smaller. This is confirmed by the comparison between the bow and stern in Figure 35. This aspect needs attention in the further optimisation of the hull in the future.

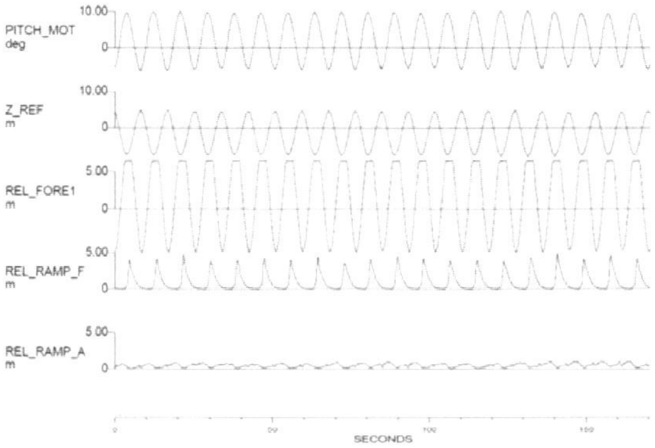

Figure 34 Time traces in a regular wave of H=3.0m without PTO modelling

Figure 35 Difference in green water flow at the bow and stern

Figure 36 shows the Pitch motion RAO in irregular waves (Tp=8.5s) with increasing wave heights of Hs=1, 2 and 3m.

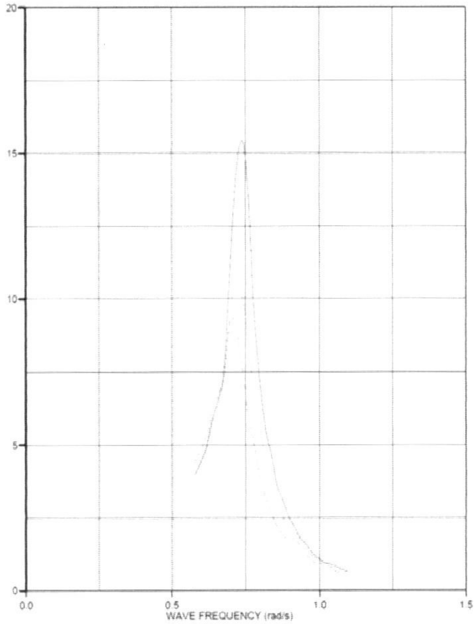

Figure 36 Pitch motion RAO in irregular wave (Tp=8.5s) with Hs=1m (solid), Hs=2m(dotted) and Hs=3m (dashed)

The RAO in low waves of 1.0m (above 15 degrees/m) confirms the predicted very large pitch motions. However, the motions are clearly non-linear: the RAOs decrease rapidly with increasing wave height, most probably due to the effect of the green water on the deck. The reduction in the pitch motions can be a measure for (part of) the amount of energy present in the green water on the deck. This needs further investigation in the future.

There were also a few more practical observations, that need careful engineering attention in the future development of the concept:

- The model in the basin needed an additional guiding wheel in front of the bow (below the pulley at deck level) to keep the wire from running off the pulley when transverse motions occurred during the tests.
- There was significant mechanical wear and tear of the wire due to the constant motions over the pulleys. This is a typical wave energy converter problem: instead of reducing the motions as is done in normal offshore engineering, the maximised motions and loads result in a significant wear and tear. Presently less sensitive mooring and PTO driving mechanisms are under investigation as a result.

CONCLUSIONS

The paper presented the initial investigations into the Green Water Concept for wave energy conversion, using on the maximisation of motions and green water. Based on the results presented, the following initial conclusions seem justified:

- The Green Water Concept combines aspects of 'overtopping', 'heaving' and 'pitching' concepts for wave energy conversion, but also adds specific aspects such as the use of green water.
- An increased draft of the structure increases the pitch motions and relative wave motions. This can be explained by the fact that the pitch excitation reduces less with the increasing draft than the wave damping. A reduced wave damping is attractive as this implies that less energy is dissipated by the wave radiation and can be used for actual wave energy conversion.
- By changing the inertia (radius of gyration) of the structure, the natural pitch period can be tuned to the frequency at which the highest wave exciting pitch moment occurs. This is important as it allows us to adjust the 'Inverse concept' to the period of the incoming waves. At resonance the spring terms in the equation of motion are cancelling the inertia terms, which results in a maximum response. The actual motion response is based on the balance between the wave excitation and the damping. In the present simulations only the wave damping is considered. In the future also the damping as a result of the electrical generator has to be taken into account.
- The iVOF simulations have shown that the principle of concentrated green water flow over the centreline of a structure can be used to fill a reservoir.
- The tests confirmed the high pitch motions and relative wave motions in low waves as predicted in the simulations. Also the concentrated high velocity green water jet along the length of the structure (needed to fill the reservoir in the centre) was clearly observed. With a wave height of only 3.0m (~1.5m single amplitude), there is already approximately 4.5m green water height on centre of the deck.
- The motions are clearly non-linear: the pitch motion RAOs decrease rapidly with increasing wave height, most probably due to the effect of the green water on the deck.
- The green water that just did not reach the reservoir resulted in a returning flow on the ramp that impacted on the incoming green water flow of the next wave cycle. This affected the effective

flow of green water onto the ramp significantly. This was not identified during the simulations as these considered only one flow event on a fixed structure. This showed the value of the model testing for the evaluation of a concept. There are options to solve this problem, like a double ramp.

The accurate modelling of a PTO is an important aspect in testing of wave energy conversion concepts numerically and in a wave tank: at the moment that energy is converted into electricity in the PTO, the hydrodynamic behaviour of the structure is changing. This will be discussed in detail in [11].

REFERENCES

[1] Cruz, J. (editor), *'Ocean Wave Energy'*, Springer, 2008, ISBN 978-3-540-74894-6

[2] Hermans, J., *'Energie survival gids, Inzicht in energie en uitzicht voor de toekomst'* (in Dutch), 2008

[3] Michael E. McCormick, *'Ocean Wave Energy Conversion'*, John Wiley & Sons, 1981

[4] Falnes, J., *'Ocean Waves and Oscillating Systems'*, Cambridge University Press, 2004

[5] Buchner, B. and Jaouen, F., 'Inverse Concept: Wave Energy Generation By Motion And Green Water Maximisation', OMAE2009-79579, OMAE 2009, Hawaii.

[6] Buchner, B.: *'Green water on ship-type offshore structures'*, PhD thesis, Delft University of Technology, 2002.

[7] 'Ocean Energy Conversion in Europe, Recent advancements and prospects', Published in the framework of the "Co-ordinated Action on Ocean Energy" (http://www.ca-oe.org) EU project under FP6 Priority: Renewable Energy Technologies, Centre for Renewable Energy Sources, 2006

[8] Fekken, G., Veldman, A.E.P. and Buchner, B.: "*Simulation of green-water loading using the Navier-Stokes equations*", Proceedings 7th Intern. Conf. on Numerical Ship Hydrodynamics, Nantes, 1999.

[9] Kleefsman, K.M.T. and Veldman, A.E.P.: "*An improved {V}olume of {F}luid i{VOF} method for wave impact type problems*", Proceedings of OMAE-FPSO 2004.

[10] Johannessen, T.B., Haver, S., Bunnik, T.H.J and Buchner, B.: *"Extreme Wave Effects on Deep Water TLPs Lessons Learned from the Snorre A Model Tests",* DOT Conference, Houston, 2006.

[11] Buchner, B. Van der Schaaf, H and Hoefakker, C.M., 'Pilot Model Tests on the 'Green Water Concept' for Wave Energy Conversion with Model Scale Power Take Off (PTO) Modelling', OMAE2010-20484, June 6-11 2010, Shanghai

PERFORMANCE PREDICTION AND DESIGN OF MARINE CURRENT TURBINES

W. Xu and S. A. Kinnas (M)
Ocean Engineering Group, Department of Civil, Architectural and Environmental Engineering, The University of Texas at Austin, Austin, Texas, USA

ABSTRACT

In this paper, the performance of a horizontal axis marine current turbine is predicted by three numerical methods. The predictions are compared with the experimental measurements for the same turbine model. A fully unsteady wake alignment is utilized in one of the methods in order to model the realistic wake geometry of the turbine. A database-searching design method is proposed to improve the turbine performance by coupling a nonlinear optimization method with one of the numerical analysis methods.

KEY WORDS

Marine Current Turbine; Boundary Element Methods; Vortex Lattice Methods; CFD; Unsteady wake alignment; Design via Optimization

INTRODUCTION

Renewable energy has become more and more important due to the global warming and limited fossil fuels. Tidal current energy is a potential renewable energy source for industry and commerce. Therefore, there is a growing interest in utilizing tidal current turbines for the extraction of energy from tidal current resources. Cavitation is the sudden formation and collapse of low-pressure bubbles in the flow around the turbine blade. It is undesirable because it produces extensive erosion of the rotating blades, additional noise from the vibrations, and a significant reduction of efficiency because it distorts the flow pattern. The ability to predict the hydrodynamic performance of a tidal current turbine and to reduce or eliminate the cavitation are essential to the design and utilization of such a system.

Experimental measurements for a 3-bladed horizontal tidal turbine were carried out in both cavitation tunnel and towing tank by Bahaj et al. (2007). Several numerical methods were used for the prediction of the performance of the same turbine model. Batten et al. (2008) used the blade element method to predict the overall performance of the turbine by combining the performances of a number of elemental blade sections. Young et al. (2009) and Baltazar et al. (2009) examined the performance of the same turbine model using a boundary element method and considered the 3-dimentional effects and expansive effects of the wake. Carcangiu (2008) applied FLUENT to analyze a wind turbine, which could also be used for the analysis of the mentioned tidal turbine model.

A turbine is basically a propeller operating in a reverse rotational direction, and most of the theories and methods originally developed for propellers can be applied to turbines. Mishima and Kinnas (1997) proposed a numerical optimization technique (CAVOPT-3D) to design cavitating propellers, in which the cubic B-spline polygon net based blade mean camber was modified at every iterative step to optimize the performance of the propeller. Kinnas et al. (2005) presented a database-searching design method, named CAVOPT-BASE, which was developed for searching the optimal propeller geometry in a performance database. Epps et al. (2009) presented a preliminary design method, using lifting line theory, for propellers and turbines.

In this work, a lower order potential based panel method (PROPCAV), a vortex lattice theory based method (MPUF-3A) and a fully viscous commercial

RANS solver (FLUENT) are used to predict the performance of the tidal current turbine studied in Bahaj et al. (2007). In MPUF-3A, the fully unsteady wake alignment is utilized to improve the prediction and to capture the expansive effects and roll up phenomenon of the wake. The cavity pattern of the turbine subject to uniform inflow is predicted by PROPCAV. The numerical results are compared with the experimental measurements. The CAVOPT-BASE was modified to work for turbines to search for the optimal geometry in the PROPCAV-based geometry database. The turbine geometry in the experiment is considered as the basic geometry. Pitch and chord length are the variables for the database. Design examples for fully wetted turbines and cavitating turbines are presented. The cavitating cases are achieved by setting a constraint which requires the minimum pressure on the blades to be larger than the vapor pressure.

METHODOLOGY

Boundary Element Method (PROPCAV)

Assume the fluid is incompressible and inviscid, and thus the flow around the turbine can be regarded as potential flow. The total flow velocity relative to the turbine fixed coordinate system can be expressed as follows:

$$\vec{q}_t(x,y,z,t) = \vec{q}_{in}(x,y,z,t) + \nabla\phi(x,y,z,t) \tag{1}$$

where $q_t(x,y,z,t)$ is the total velocity, $q_{in}(x,y,z,t)$ is the inflow velocity and $\phi(x,y,z,t)$ is the perturbation potential. The perturbation potential satisfies the Laplace's equation in the fluid domain.

$$\nabla^2\phi(x,y,z,t) = 0 \tag{2}$$

By applying Green's third identity with respect to the perturbation potential ϕ at any time, the following integral equation can be reached:

$$\begin{aligned}2\pi\phi(\bar{x},t) = &\iint_{S_B+S_C}\left[\phi_q(\bar{x},t)\frac{\partial G(p;q)}{\partial n_q} - G(p;q)\frac{\partial\phi_q(\bar{x},t)}{\partial n_q}\right]ds \\ &+\iint_{S_W}\Delta\phi_W(\bar{x},t)\frac{\partial G(p;q)}{\partial n_q}ds \\ &+\iint_{S_T}\phi_q(\bar{x},t)\frac{\partial G(p;q)}{\partial n_q}ds\end{aligned} \tag{3}$$

The subscript q indicates the variable points in the integration; n_q is the unit normal vector of the body surface or the wake surface. $G(p,q)$ is the Green's function and $\Delta\phi_w$ is the potential jump at the trailing edge of blades. In addition, the kinematic boundary condition on the wetted foils is that only a tangential velocity exists on the body surface, as follows:

$$\frac{\partial\phi}{\partial n} = \nabla\phi\cdot\vec{n} = -\vec{q}_{in}\cdot\vec{n} \tag{4}$$

The dynamic boundary condition on cavity surface states that the pressure there is equal to the vapor pressure. In PROPCAV, the effects of viscosity on the blades are taken into consideration via the application over the wetted part of the blades of a prescribed friction coefficient. More details of the method can be found in Fine (1992); Fine and Kinnas (1993); Kinnas and Fine (1992).

Vortex lattice method (MPUF-3A)

In Vortex lattice method (VLM), the blade loading and blade or cavity thickness are represented by the discrete line vortices and line sources. The vortices and sources are placed on the mean camber surface and its trailing wake surface.

The discrete vortex segment forms a circulation horseshoe with constant strength Γ and the discrete source segment represents the effects of blade and cavity thickness. The kinematic and dynamic boundary conditions at control points are applied to determine the unknown strength Γ and the cavity source strength Q_C. The kinematic boundary condition requires the flow to be tangent to the mean camber surface:

$$\sum_{\Gamma}\Gamma\vec{v}_\Gamma\cdot\vec{n} + \sum_{Q_B}Q_B\vec{v}_Q\cdot\vec{n} + \sum_{Q_C}Q_C\vec{v}_Q\cdot\vec{n} = \vec{q}_T\cdot\vec{n} \tag{5}$$

where $\vec{q}_T$ is the inflow velocity, $\vec{n}$ is the unit normal vector, and Γ is the constant strength of bound vortex. Q_B and Q_C are the line sources corresponding to the blade and cavity thickness, respectively. $\vec{v}_Q$ and $\vec{v}_\Gamma$ are the velocity vectors induced by each unit strength of line source and vortex loop. In VLM, the dynamic boundary condition and friction effect are applied in the same way as that of Boundary Element Method. More details of this method can be found in Lee (1979).

Fully unsteady wake alignment

The wake alignment is essential to the prediction of the performance of the marine current turbine. In the VLM, the fully unsteady wake alignment is used to automatically adjust the wake geometry of the turbine. The unknown singularities are firstly solved based on the unaligned wake geometry. Locations of a new wake geometry are determined by applying the force-free condition on wake. The control points will move along the total velocity each iterative step until convergence is reached. The vortex core at the tip of blades is not included in this work. More details of the fully unsteady wake alignment can be found in Lee and Kinnas (2005). One application of the fully unsteady wake alignment in MPUF-3A for propellers is presented in Lei and Kinnas (2009).

RANS Solver (FLUENT)

In this work, the viscous flow field around the marine current turbine is analyzed using FLUENT 6.3, which solved the continuity and the momentum equations by employing a cell-centered finite volume method. Pressure–velocity coupling is achieved by using SIMPLE algorithm. The *k-ω* model is used for turbulent flow modeling. The moving reference frame is used and the periodic condition is applied to reduce the cell number. Thus only 1/3 of the total domain is used and the total mesh number for this case is 930,000. The analysis is performed on a 16 nodes cluster (two 2.4 GHZ quad-core 64-bit Intel Xeon processors with 16 GB Memory per node). The convergence criteria for the residuals are equal to 1×10^{-6}. It takes about 5 hours (16 parallel cores) to converge.

Database-searching design method (CAVOPT-BASE)

CAVOPT-BASE uses a constrained nonlinear optimization method to search for optimal turbine geometry in a performance database. The database is constructed by analyzing turbines with different geometries. At the first step, a base turbine is selected from well-known turbines, and then a family of the base turbine is generated by changing the geometric parameters.

$$(P/D)_{design} = (P/D)_{base} \times x_1$$
$$(c/D)_{design} = (c/D)_{base} \times x_2$$
$$(Y_{p,s}/c)_{design} = (Y_{p,s}/c)_{base} \times x_3 \qquad (6)$$

where *P/D*, *c/D*, and $Y_{p,s}/c$ represent normalized pitch, chord and blade offsets of the pressure or suction side, respectively. *D* is the diameter of the turbine, x_1, x_2, and x_3 are three multipliers corresponding to three design parameters. Lower bound and upper bound are specified for each multiplier, and the density of the database can be determined by the number of intervals between these two bounds. The constrained nonlinear optimization problem in this work has the following form:

$$\text{maximize } K_Q(x)$$
$$\text{subject to } C_{P\min}(x) \le CPMIN$$
$$x_i^{LB} \le x_i \le x_i^{UB} \qquad i = 1,2,3 \qquad (7)$$

where x_i^{LB} and x_i^{UB} are the lower and upper bound for each multiplier x_i, $K_Q=Q/\rho n^2D^5$ is the torque coefficient, $C_{Pmin}=\max\{-Cp\}$ is the minimum of pressure coefficient $C_P=(P-P_0)/(0.5\rho n^2D^2)$, *Q* is the torque, ρ is the fluid density, P_0 is the far upstream pressure at the turbine shaft, *P* is the pressure on blade, *CPMIN* is the maximum allowed C_{Pmin} which corresponds to the cavitation number $\sigma_n=(P_0-P_v)/(0.5\rho n^2D^2)$, P_v is the vapor pressure. The torque coefficient relates to the power coefficient by Cpow=16TSR/πK_Q. More design parameters and more constraints on the maximization problem can be added for other design cases. The details of this method can be found in Deng (2005) and Kinnas et al. (2005).

It should be noted that in order to keep the maximum thickness the same among different designs and not compromise the structural integrity of the turbine, we require that $x_3=1/x_2$

Therefore, only the parameters x_1 and x_2 are to be determined in the optimization.

NUMERICAL RESULTS

Comparison with experiments

The turbine geometry is given based on the experimental study of Bahaj et al (2007). The blades were developed from the profile shape of a NACA 63-8xx, together with specified chord, thickness and pitch distributions. The experiment was performed with a fixed turbine inside a 2.4m by 1.2m cavitation tunnel. In this paper, hub and lower circular part of the blades are not included in the numerical model and the axis is oriented at zero yaw angle. The hub effect is small due to thin circular connection part between the hub and blades. Numerical predictions from PROPCAV, MPUF-3A and FLUENT are compared with experimental data for a turbine with a hub pitch angle (Phi) of 25 degrees operating at different tip speed ratios (TSR=$\omega R/V$). *V* is the mean inflow velocity defined at the hub axis, $\omega=2\pi n$ is the turbine angular frequency, and $R=D/2$ is the turbine radius.

Comparisons of the predicted thrust coefficient (CT=$T/0.5\rho V^2A$) and power coefficient (Cpow=$Q\omega/0.5\rho V^3A$) are shown in Figs. 1 and 2, where $A=\pi R^2$ is the turbine sweep area, *T* is the flow generated thrust. All of these three methods give reasonable predictions for the performance of turbines. The FLUENT has a better prediction for high TSR (=8), while the results from PROPCAV and MPUF-3A are better when compare with the experimental measurements for design TSR (5≤TSR≤6). The predictions of the performance for the same turbine operating at 27 and 30 degrees hub pitch, which are not shown on these two figures, also compare well with the experimental measurements. It should be noted that the computational time for PROPCAV and MPUF-3A is less than 3 minutes on one node, which is much less than what is needed for FLUENT (5 hours) on 16 nodes.

The wake alignment has been found to be essential to the performance of the turbine. In this research, blade pitch wake alignment is utilized for PROPCAV, in which we specify the ultimate wake radius as 1.2R and the wake expansion angle as 30 degrees. In MPUF-3A, the unsteady wake alignment (Lei and Kinnas 2009) is utilized. The wake geometry is determined automatically by applying the force free condition. The initial wake geometry is produced without any alignment. In FLUENT, the wake geometry is presented by streamlines. A full domain analysis case is carried out in order to investigate the wake geometry. The wake geometries from these two potential flow solvers and FLUENT are shown in Figs. 3 ~5. The wake expansion effect can be observed in the PROPCAV by using geometric wake alignment (Fig. 3). In MPUF-3A (Fig. 4), both radial expansion and tip roll-up of the wake can be observed when the convergence of the unsteady wake alignment is reached. It takes up to 5 or 6 iterations for the force free condition to be satisfied. The FLUENT result in Fig. 5 also indicates the tip roll up phenomenon (Fig. 5(b)) and wake radial expansion effect (Fig. 5(c)).

A cavititing case has also been investigated by PROPCAV, in order to validate the predicted cavitation pattern. A turbine model operating at TSR=7.5 with 25 degrees hub pitch and 0 yaw is analyzed. The static pressure of the cavitation tunnel is P_0=23,000Pa and

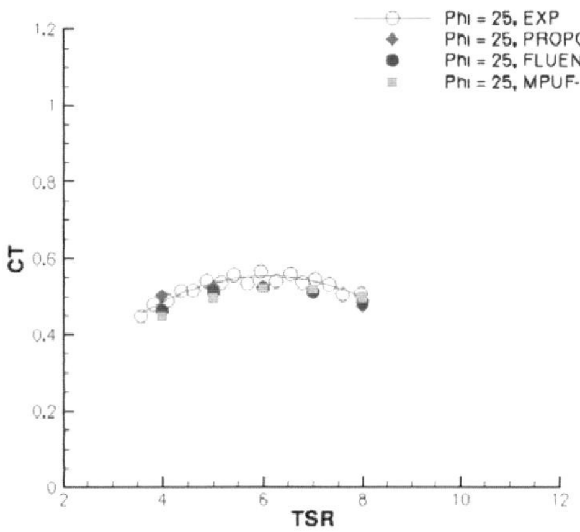

Fig. 1 Comparison of thrust coefficient (CT) from different numerical methods with the experimental measurements

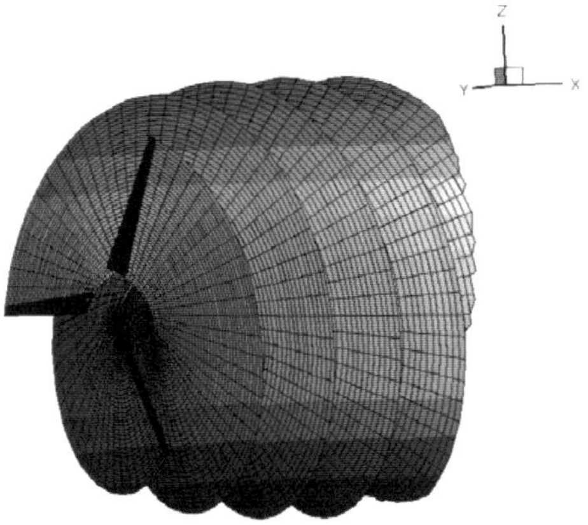

Fig. 3 Wake geometry from PROPCAV by using geometric pitch alignment

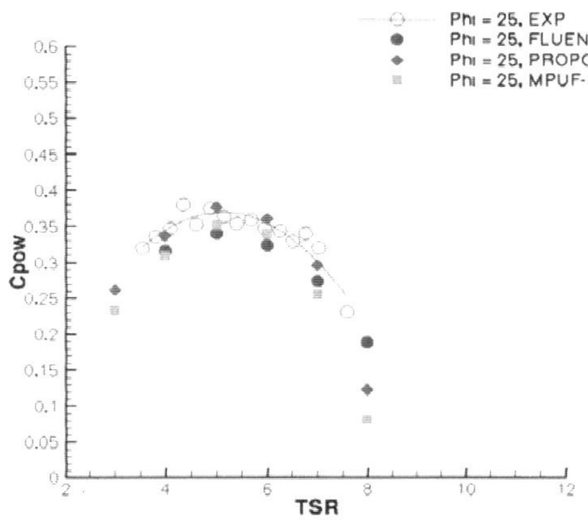

Fig. 2 Comparison of power coefficient (Cpow) from different numerical methods with the experimental measurements

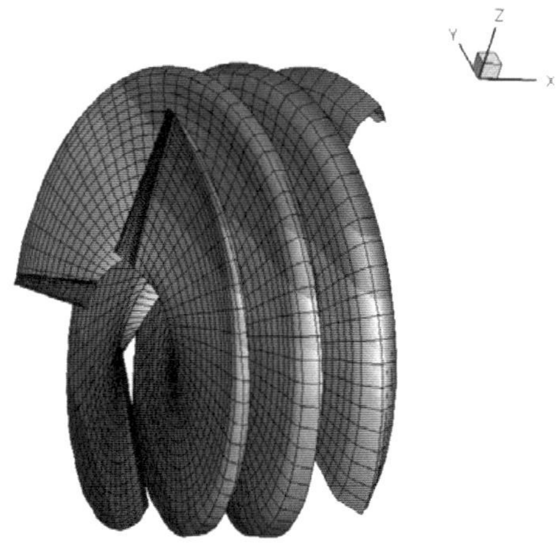

Fig. 4 Wake geometry from MPUF-3A by using fully unsteady wake alignment

the vapor pressure of water at 10^0C is P_v=1,230*Pa*. Thus the cavitation number defined at the axis of rotation is σ_n=3.9 (*CPMIN*=3.9). The predicted cavity pattern, as well as the photograph of experimental observation, is shown in Fig. 6.

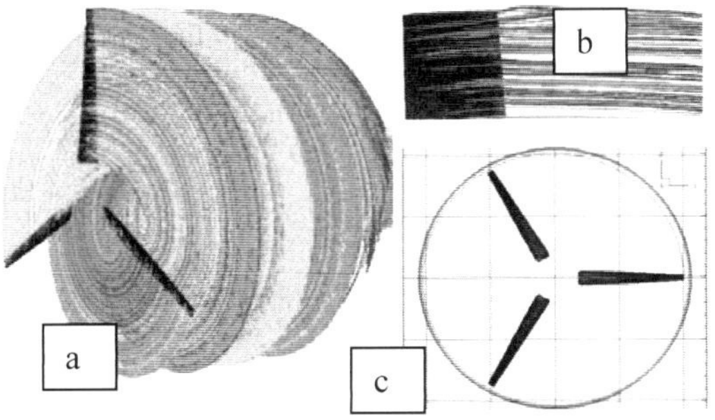

Fig. 5 Wake geometry from FLUENT. (a) wake visualized by streamlines, (b) tip roll up phenomenon, (c) expansion effects visualized by the tip streamline.

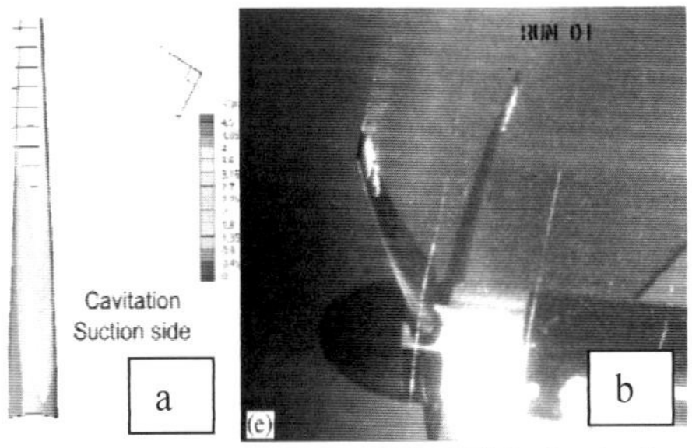

Fig. 6 The cavity pattern-TSR=7.5, 25° hub pitch. (a) from PROPCAV, (b) from Bahaj et al. (2007)- Note the turbine in the experiment is left handed while the calculation is using a right handed rotation convention.

Example applications and implications

Fully wetted turbine design

CAVOPT-BASE method, which couples PROPCAV and a nonlinear optimization method, is proposed for the rotor design in this study. A 20m diameter tidal turbine, suggested by Blunden and Bahaj (2007), is used as an example here. Based on the power coefficient curve (Fig. 2) and the given tidal velocity data (Batten et al. 2007), the suitable design TSR and design tidal speed range from 4 to 7 and 2m/s to 3m/s, respectively. For the fully wetted design case, *CPMIN* is required to be large enough to inactivate the pressure constraint. We assume TSR=5.0 and the tidal velocity U=2.5m/s for this case. Thus the turbine rotational frequency is n=0.24rev/s. The CT and Cpow for the original geometry are 0.55 and 0.39. The pitch and chord length are selected as the two geometric parameters to produce a performance database. The thickness of the blade is kept the same for all geometries. The two multipliers (x_1 and x_2), which correspond to pitch and chord length, range from 0.80 to 1.2 and 0.8 to 1.2. The design results are Cpow=0.412 and CT=0.586 when x_1=0.803 and x_2=1.006. The maximum –Cp (C_{Pmin}) for the design geometry is 7.78, which is very close to the value (7.26) of the original geometry. Both Cpow and CT increase in this design. Thus there is clearly a balance between the extra costs of the support structure (larger CT) and the extra power (larger Cpow) that can be generated. The database size is very limited in this design case, and a better geometry may be found if a larger database is used.

Cavitating turbine design

For the larger size turbines that has shallow tip immersion, cavitation may become a serious issue. The design method CAVOPT-BASE can also be used to reduce or eliminate the cavitation by adjusting the geometric parameters of a certain turbine. In this work, two cavitating cases are investigated and the same numerical turbine model is used.

Case 1

A 25 degrees hub pitch turbine operates at TSR=5.0 and cavitation number σ_n=5.0. The pressure distributions of different blade sections are shown in Fig. 7. The x coordinate is the ratio of section point locations with respect to the section chord length. Thrust coefficient and power coefficient are equal to CT=0.486 and Cpow=0.317. As is shown in Fig. 7, cavitation happens on part of the blade. The cavitation pattern for this case is close to what is shown in Fig. 6. The pitch and chord length are also selected as the two geometric parameters to produce a performance database. The ranges of the two multipliers are $1.0 \le x_1 \le 1.2$ and $0.8 \le x_2 \le 1.5$. In order to eliminate the cavitation, the inequality constraint $C_{Pmin} \le 5.0$ is applied in the optimization problem. The design result from CAVOPT-BASE is x_1=1.006 and x_2=1.351. In this database, the optimal geometry, which satisfies the constraint, is the one with a longer chord length and almost the same pitch. The thrust and power coefficients for the design case are CT=0.516 and Cpow=0.371. The pressure distributions for different sections are shown in Fig. 8. It is noted that cavitation is successfully eliminated and the power coefficient increases as well. Nevertheless, the thrust coefficient also increases which can also lead to structural problem for the turbine. There is clearly another balance

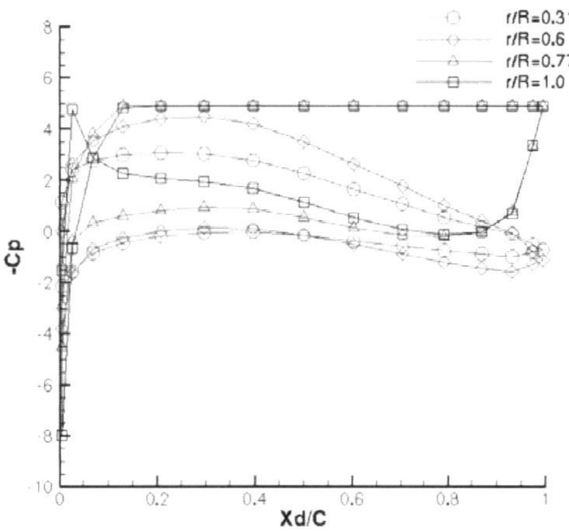

Fig. 7 Pressure distribution of the original cavitation case at different radii for case 1

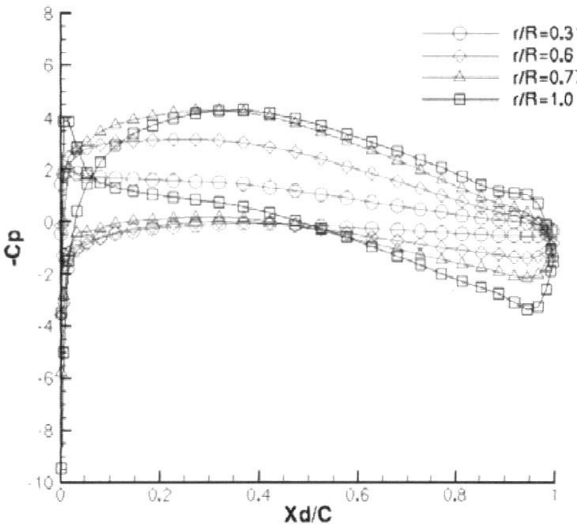

Fig. 8 Pressure distribution of the design fully wetted case at different radii for case 1

between the extra costs of support structure (larger CT) and the benefits from the non-cavitation working condition and extra power (larger Cpow).

Case 2

A turbine with 40m diameter and 25 degrees hub pitch, operating at TSR=7.5 is analyzed in this case. The tip immersion for this turbine is 5 meters and the tidal velocity is 3m/s. Thus the rotational frequency is n=0.179rev/s.

The cavitation number is σ_n=11.4 and the Froude number is $Fr=n^2D/g$=0.1308. The cavity patterns at several angular positions are shown in Fig. 9. The thrust coefficient and power coefficient are CT=0.497

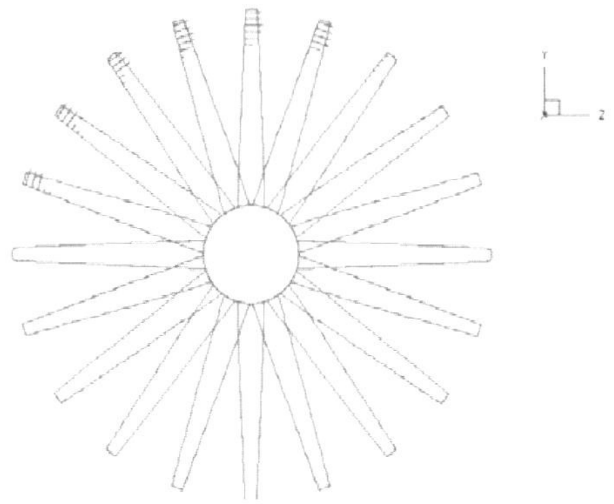

Fig. 9 Projected Blade geometry and cavity patterns at various angular positions for the original turbine (case 2)

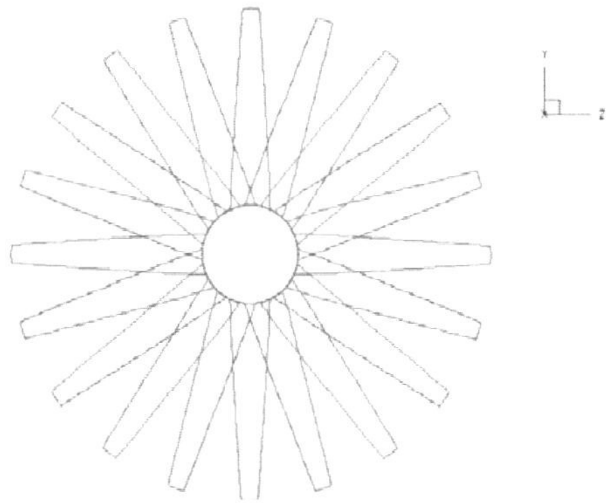

Fig. 10 Projected Blade geometry and cavity patterns at various angular positions for the design turbine (case 2)

and Cpow=0.153. The database is produced by selecting pitch and chord length as geometric parameters. The ranges of the two multipliers are changed to be $0.8 \le x_1 \le 1.2$ and $0.8 \le x_2 \le 1.5$, respectively. The fully wetted solution cannot be found in this database, which means we can reduce the cavitation rather than eliminate it by searching this database. The projected blade geometry and the cavity pattern of the optimal turbine, which reduce the cavity most when x_1=0.8 and x_2=1.445, are shown in Fig. 10. The thrust and power coefficients which correspond to the design geometry are CT=0.697 and Cpow=0.342. By comparing Fig. 9 with Fig. 10, we see that the cavitation is greatly reduced by decreasing the pitch and increasing the chord length, while keep the thickness of the blade be the same. All design conditions and results are tabulated in Table1.

Table 1. Three design cases by CAVOPT-BASE

	Fully wetted case	**Cavitating case 1**	**Cavitating case2**
Design Parameters	P/D c/D	P/D c/D	P/D c/D
Multipliers' Range	$0.8 \le x_1 \le 1.2$ $0.8 \le x_2 \le 1.2$	$1.0 \le x_1 \le 1.2$ $0.8 \le x_2 \le 1.5$	$0.8 \le x_1 \le 1.2$ $0.8 \le x_2 \le 1.5$
TSR	5.0	5.0	7.5
Design Multipliers	x_1=0.803 x_2=1.006	x_1=1.006 x_2=1.351	x_1=0.800 x_2=1.445
(*CPMIN*) Design σ_n	N/A	5.0	11.2
Original KT KQ	0.550 0.390	0.486 0.317	0.497 0.153
Design KT KQ	0.586 0.412	0.516 0.371	0.697 0.342

CONCLUSIONS

Three numerical analysis methods, panel method PROPCAV, vortex lattice method MPUF-3A and RANS solver FLUENT are applied to investigate the performance of tidal current turbines. The results from these methods are all consistent with the experimental measurements within a reasonable range of difference. However, it takes up to 5 hours for FLUENT to finish the fully wetted analysis, while 3 minutes are enough for both PROPCAV and MPUF-3A.

A database-searching design method CAVOPT-BASE is applied to maximize the power coefficient and reduce or eliminate the cavitation. Note that the increase of the thrust coefficient may lead to structural problems, which can be solved by adding an additional constraint that specifies the maximum thrust coefficient for the design case. The design method can be used in the rotor design procedure by providing more geometry options and meeting additional design requirements by adding more equality or inequality constraints in the optimization method. The effectiveness of the design method is naturally based on the quality of the results from the performance analysis and the size of the database. CAVOPT-3D will be used in the future, which can determine the blade geometry automatically and no database is needed.

ACKNOWLEDGEMENTS

The authors wish to thank Dr. Yi-Hsiang Yu, Dr. Vimal Vinayan, Shu-Hao Chang and Lei He of the Ocean Engineering Group at UT Austin for their valuable help with running the involved codes in this paper.

REFERENCES

[1] BAHAJ, A., MOLLAND, A., CHAPLIN, J., AND BATTEN, W. "Power and thrust measurements of marine current turbines under various hydrodynamic flow conditions in a cavitation tunnel and a towing tank". *Renewable Energy*, 32(3): (2007): 407 – 426.

[2] BALTAZAR, J., FALCAO DE CAMPOS, J.A.C. "Unsteady analysis of a horizontal axis marine current turbine in yawed inflow conditions with a panel method," First International Symposium on Marine Propulsors, Trondheim, Norway, (2009)

[3] BATTEN, W., BAHAJ, A., MOLLAND, A., CHAPLIN, J. "Experimentally validated numerical method for the hydrodynamic design of horizontal axis tidal turbines", *Ocean Engineering*, 34 (2007): 1013-1020

[4] BATTEN, W., BAHAJ, A., MOLLAND, A., AND CHAPLIN, J. (2008) "The prediction of the hydrodynamic performance of marine current turbines". *Renewable Energy*, 33 (2008): 1085 – 1096

[5] BLUNDEN LS, BAHAJ AS. "Tidal energy resource assessment for tidal stream generators", *Proc Inst Mech Eng*, Part A. J Power Energy; 221:2 (2007): 137-46

[6] CARCANGIU, C. "CFD-RANS Study of Horizontal Axis Wind Turbines," Phd Thesis, Dipartimento di Ingegneria Meccanica, DIMeCa, Universita degli Studi di Cagliari, 2008

[7] DENG, Y. "Performance Database Interpolation and Constrained Nonlinear Optimization Applied to Propulsor Blade Design," Master Thesis, the University of Texas at Austin, Ocean Engineering Group, Department of Civil, Architectural & Environmental Engineering, 2005

[8] EPPS, B., STANWAY, M., AND KIMBALL, R. OpenProp: An open-source design tool for propellers and turbines, 2009

[9] FINE, N. "Nonlinear Analysis of Cavitating Propellers in Nonuniform Flow," PhD thesis, Department of Ocean Engineering, MIT.

[10] FINE, N. E. AND KINNAS, S. A. "A boundary element method for the analysis ofthe flow around 3-d cavitating hydrofoils," *Journal of Ship Research*, 37 (1993): 213–224

[11] HE, L. AND KINNAS, S. "A Method for the Prediction of Rotor/Stator Interaction-Applications to 2-D Hydrofoils and Propeller," *Propeller and*

Shafting Conference, Williamsburg, VA, (2009)

[12] KINNAS, S. A. AND FINE, N. E. A nonlinear boundary element method for theanalysis of unsteady propeller sheet cavitation. In Nineteenth ONR Symposium on Naval Hydrodynamics, pages 717–737 (1992)

[13] KINNAS, S., LEE, H., GU, H., AND DENG, Y. "Prediction of Performance and Design via Optimization of Ducted Propellers Subject to Non-axisymmetric Inflows", *Trans. Society of Naval Architects and Marine Engineers*, 113 (2005)

[14] LEE, CS "Prediction of Steady and Unsteady Performance of Marine Propellers with or without Cavitation by Numerical Lifting Surface Theory," PhD Thesis, MIT, Department of Ocean Engineering, (1979)

[15] LEE, H. AND KINNAS, S. "Fully Unsteady Wake Alignment for Propellers in Non-axisymmetric Flows," *Journal of Ship Research*, 49:3 (2005): 176-190

[16] MISHIMA, S, KINNAS, S. "Application of a numerical optimization technique to the design of cavitating propellers in nonuniform flow". *Journal of Ship Research*, 41:2 (1997): 93-107

[17] YOUNG, Y., MOTLEY, M., AND YEUNG, R. "Hydroelastic Response of Wind or Tidal Turbines," *The ASME 28th International Conference on Ocean, Offshore and Arctic Engineering*, Honolulu, Hawaii, (2009)

SESSION C PAPERS

(This page intentionally left blank.)

EXPERIENCES IN ROPE DESIGN FOR OFFSHORE RENEWABLE ENERGY PROJECTS

Karel Devos, Carina De Plukker, Peter Van den Berghe
Bexco Ropes

ABSTRACT

In this paper an introduction is given into the rising offshore renewable energy market. It's pointed out that dynamic stiffness is a key requirement for mooring ropes in these systems. Axial stiffness is explained, as well as the difference between static and dynamic load regime, from the material morphology point of view. For dynamic stiffness the influencing parameters mean load, load range and frequency are distinguished and it is mentioned that according to previous research in a real wave situation the parameters load range and frequency can be ignored.

A first example case discusses a renewable wave energy device, where the rope construction and rope layout was completely custom designed to fit the application. Studies were done into permanent Chinese Fingers pulling slings and full scale pulling eyes. These special slings assist in performing the installation and de-installation(recuperation) most efficiently.

For a specific and recent offshore renewable wind energy project, a comparison is made between different high modulus fiber materials, to which material is the most economically suitable. The application requires a high dynamic stiffness and low creep elongation, besides 25 year fatigue lifetime. Different high modulus fiber materials like Vectran UM, Twaron 2200, Dyneema SK78 and Carbon fiber (TohoTenax STS40) are compared. The main conclusion is that in the range of service loads considered, many materials can be competitive, and therefore future work should test the suitability and cost efficiency of these materials on subrope level.

A. INTRODUCTION

The renewable energy market demand.

The last decade, companies in the renewable energy market are fighting hard to come up first with the most cost-efficient concept to harness energy from natural resources and to fund a full scale prototype test in an offshore test site. This test is a critical factor in proving the survivability and performance of any system. Efforts are focused on decreasing the cost per kW hour generated electricity, both for of the energy device, as on the mooring system of the device. For example for floating windmills the mooring system represents up to 13% of the total costs compared to a standard non-floating windmill where the costs of the foundations can be up to 60%. Designing the devices and their mooring systems to 100-year storm conditions, enlarges the potential destinations for these devices, but the challenge is to maintain a high conversion efficiency.

Once the basic project specifications are determined it is recognised that there is often still a lack of experience with synthetic ropes, and therefore a need for good technical partners. In these projects, the rope producer may be involved very early in the project to help determine the rope design specifications. At the stage of the prototype market, confidentiality between client and producer is key and the starting point for an intensive cooperation.

The international community is heavily supporting this new market, e.g. the carbon trust in UK and many other government initiatives. Pressure is put to install these new power stations in the years to come. Demonstrating the viability of the system and certification to be compliant with international standards needs to go fast. Companies in the renewable energy industry still prefer independent consultants with international reputation, as governments that need to be convinced of these systems span across the world.

Axial stiffness key requirement

Typically, the requirements for moorings tendons for renewable (wind & wave) energy projects are comparable to deepwater mooring systems for oil platforms & FPSO's. However, besides MBL, low creep elongation and fatigue lifetime ranging from 15 to 25years, also axial stiffness tends to be a key requirement in these applications. Why this is the case has to do with the specifics of the mooring systems and is not part of the scope of this paper. But, we will see later on that the stiffness requirement will play a dominant role in the rope design.

B. AXIAL STIFFNESS OF A MOORING TENDON

Stiffness of a mooring tendon

When a tension member is being loaded, a certain elongation is being enforced. The stress in the material is the resistance of the material to this elongation. The level of this stress per unit elongation is a material constant namely the stiffness (Young's modulus E).

In a load-elongation curve, stiffness is the stress reaction (or tension) per elongation (= Δ load / Δ elongation). Hence stiffness is equal to the tangent of the angle of the curve. The Young's modulus E of a fibre is expressed in Gpa. Stiffness of a specific tension member is often given as EA, with A the total fibre surface. EA is expressed in kN or kN/m, which is the stiffness divided by the length of the member.

The larger the rope diameter, the larger the stiffness as more material will add to the total stiffness. A larger rope will react with a higher load for the same elongation. In order to compare the stiffness of ropes of different sizes, stiffness can be expressed in a non-dimensional form as parameter Kr (=EA/MBL). Kr is expressed in x MBL (times MBL).

Fibre stiffness versus rope stiffness

Figure 1 below shows the Young's modulus (Gpa; supplier documentation) of different commercially available fiber materials. The wide modulus range of carbon fibers is remarkable. Figure 2 also displays the very wide range of available strength-modulus combinations for carbon, compared to other materials.

The stiffness of a specific mooring tendon is the result of both the fiber material stiffness and the construction of the rope. Classical rope constructions may be either braided, twisted or parallel constructions based on twisted strands, see figure 3. The angle of the construction determines the efficiency, both for strength and for stiffness. The more parallel the fibers are the higher the efficiency. At the rope level, a braided or twisted construction brings in construction elongation which increases the total elongation of the tendon. Under load a new rope will be compacted (air is squeezed out), assisted by small diameter reduction of the elongated fiber. As the construction is compacted, it will also elongate (up to 5% depending on the construction type). Using this principle, elongation of a tendon can be custom designed to specific application needs. An example case is discussed further in this paper.

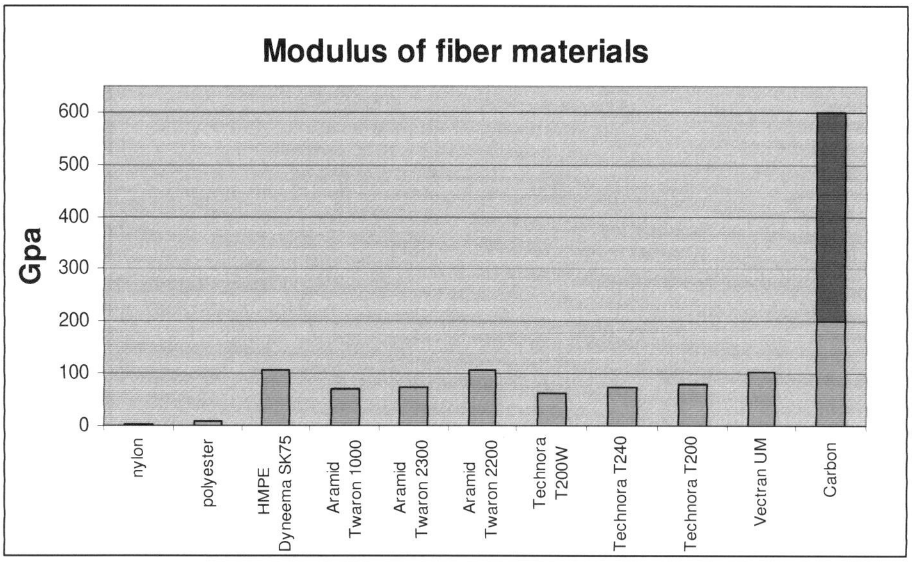

Figure 1. Young's modulus of fiber materials

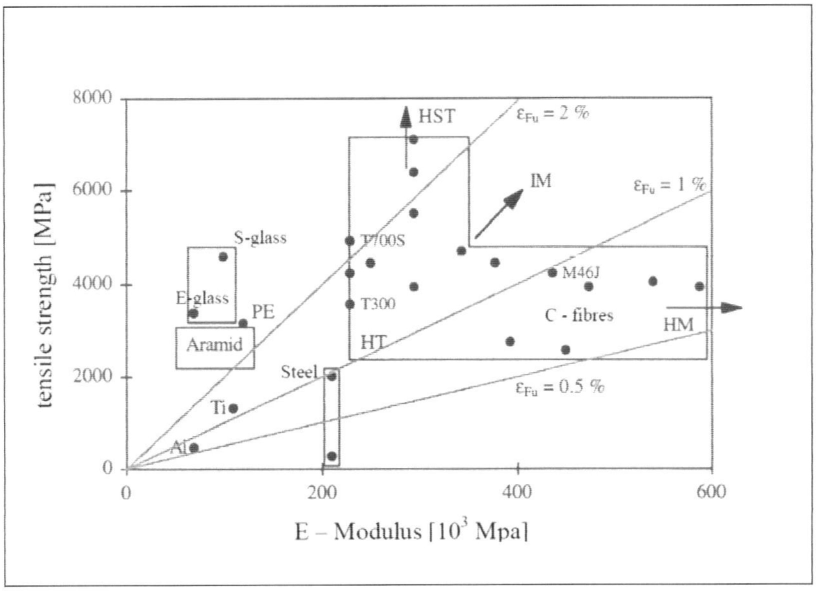

Figure 2. Strength and modulus data of available fiber materials.

Laid ropes	3-strand 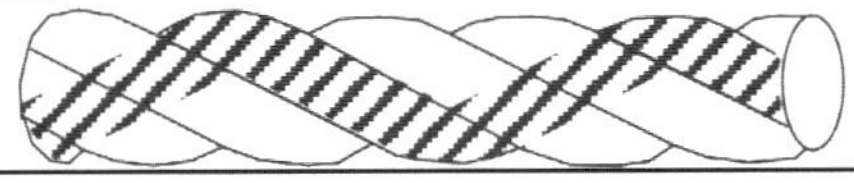	-rotate under load (small effect on strength) -line contacts between the strands -good fatigue life -good strength efficiency
	6-strand 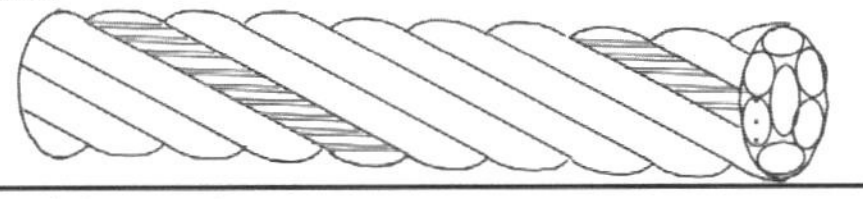	
Braided ropes	8-strand (square) 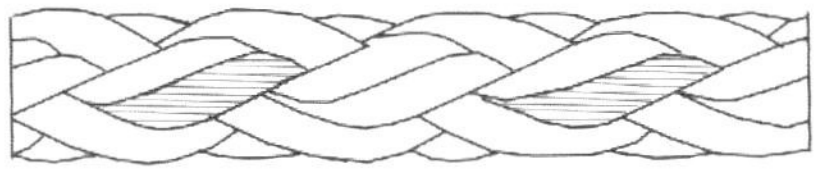	-will not rotate under load -point contacts between the strands -lesser fatigue life -higher constructional elongation -less strength efficiency -better bending flexibility
	12-strand (circular) 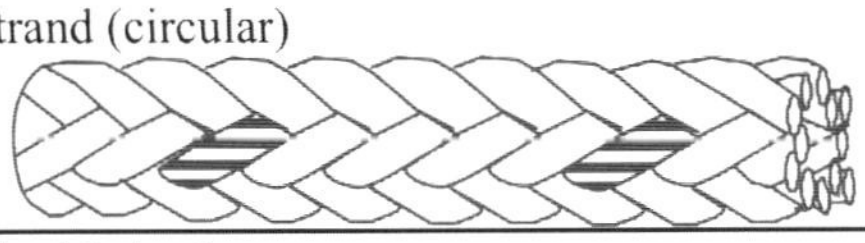	
Parallel ropes	overbraided sub ropes 	-optimal strength efficiency -left laid + right laid cores result in torque balance

Figure 3. Classical synthetic rope constructions

C. STATIC VERSUS DYNAMIC STIFFNESS

Stiffness behaviour of fiber materials - molecular & macro-molecular structure

The behaviour of synthetic fibers, especially the elongation behaviour, is very much depending on the macro-molecular structure of the material (also called 'morphology'), besides the molecular formula itself. The macro-molecular structure is the way the molecules are organized when the fibers are being produced. In this we can group the polymer materials like HMPE, Vectran, aramids and Technora. Polymer molecules per definition consist of very long carbon chains with different functional groups that determine the specific characteristics.

The morphology of polymer materials typically shows crystalline parts and non-crystalline (amorphous) parts. When a polymer like PE (basis for HMPE) crystallizes from the melt, the crystallization of the carbon chains into crystalline and amorphous parts starts from tiny nuclei and finishes into what is called a 'spherulite', see figure 4. Carbon chains 'grow' through both the crystalline and amorphous part, linking them together. From PE, HMPE is produced by aligning up to 95% of the molecules in a melted stage, maximising the crystalline part. Other high modulus and high strength fibers such as Vectran, aramids and Technora are also characterized by a high degree aligned molecules.

As can be expected, the crystalline and amorphous parts have very different properties. The crystalline part is associated with high density, high strength and rigidity, while the amorphous part is generally less dense, weaker and more easily deformed. We will show that this will play a important role in the specific stiffness character of these fiber materials.

Carbon fiber on the other hand also consist of carbon atoms, but is classified under the ceramic materials. A single such filament is a thin tube with a diameter of 5–8 micrometers and consists almost exclusively of carbon atoms. The atomic structure of carbon fiber is similar to that of graphite, consisting of sheets of carbon atoms (graphene sheets) arranged in a regular hexagonal pattern. See figure 5. Clearly the macro-molecular structure is very different, without any amorphous part, and the specific stiffness behaviour of carbon fiber will be accordingly.

As polyester was more used for station keeping of platforms and FPSO's during the '90, there was an increasing need to predict the elongation behaviour of polyester ropes. A stiffness model has been proposed by J. Flory et al, see figure 6. In this model we can differentiate between fiber material factors and constructional factors, contributing to the elongation. The material factors are the responses of the crystalline part and the amorphous part and creep elongation. The constructional factor is associated with construction

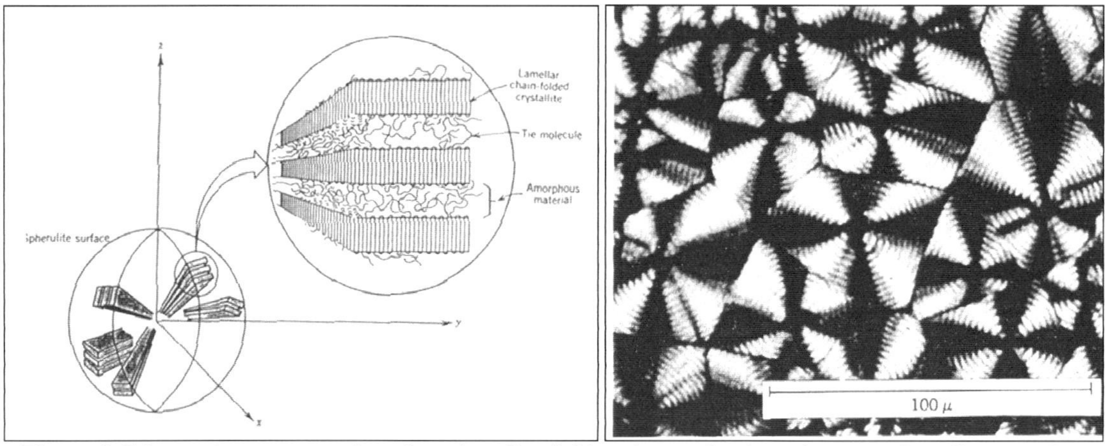

Figure 4. Schematic representation of a spherulite (left); microsopic view of many spherulites together (right).

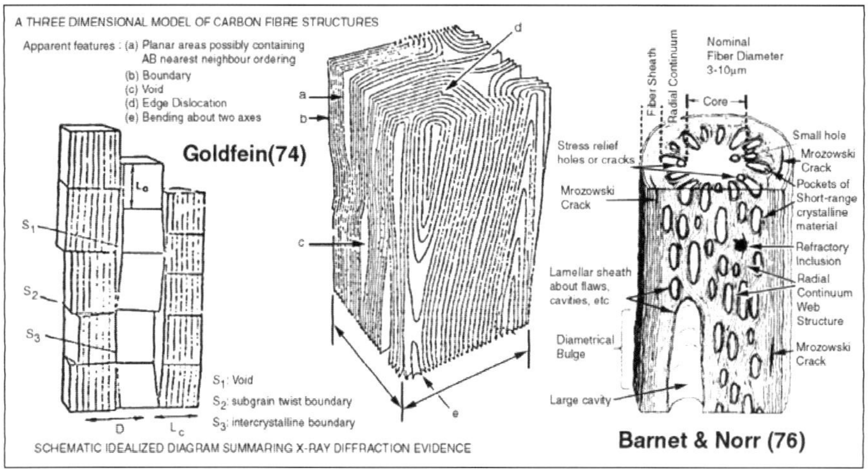

Figure 5. Macro-molecular structure of carbon fiber.

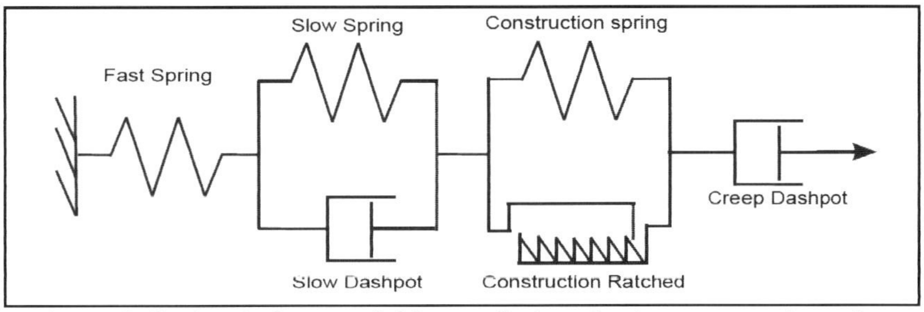

Figure 6. Spring-dashpot model for prediction of polyester rope elongation

elongation when a new rope is loaded for the first time and the construction is compacted.

Crystalline part: Fast spring. Stretch which occurs when the rope is rapidly loaded (after the rope has been initially loaded to remove construction stretch).

Amorphous part: Parallel slow spring and dashpot system. Additional stretch which occurs when the rope is slowly loaded. The slow spring's rate of stretch as well as its rate of contraction are retarded by the dashpot. The stretch which occurs when the rope is slowly loaded is the summation of the fast and slow spring stretches.

Rope construction stretch: Parallel construction spring and dashpot system. Construction stretch is permanently removed when load is initially applied to the rope. The ratchet prevents contraction of the spring. But whenever a higher tension is applied this construction spring/ratchet system stretches even further.

Material creep: Creep dashpot. This creep is typically rapid at first and decreases with time (except for HMPE, where creep is linear).

Static stiffness

Static stiffness is the stiffness of a tension member when it is loaded slowly, e.g. in a break test from 0-100%MBL in several minutes, and after bedding-in of the construction. The loading rate is slow and leaves time for both the amorphous and crystalline part to react the load. The resulting fiber stiffness is an average of the stiffness of both parts.

Dynamic stiffness

Dynamic stiffness is the stiffness response of a tension member when it is cyclically loaded. Typically for polymer materials the dynamic stiffness is higher than the static stiffness. As the amorphous part does not react fast enough to the quickly changing loading regime, it is the (stiffer) crystalline part that takes on the load, resulting in a more ridged response of the whole fiber. When cycling has started, it takes a number of cycles (around 100) before the stiffness increases from the static stiffness up to the dynamic stiffness level. For this reason dynamic stiffness tests should always run a minimum of 100 cycles. As opposed to static stiffness, the dynamic stiffness level is influenced by different parameters of the cycling regime: mean load, load range and cycling period (frequency). Mean load and frequency have a positive effect on dynamic stiffness, while load range has a negative effect.

Davies et al, determined the quantitative effects of these parameters on dynamic stiffness, for high modulus fibers Dyneema SK75, Twaron 1000 in comparison with polyester Diolen 855TN [1]. In testing the mean load effect, cycling was performed with a fixed load range of +/ 5% and 10s cycle period, and

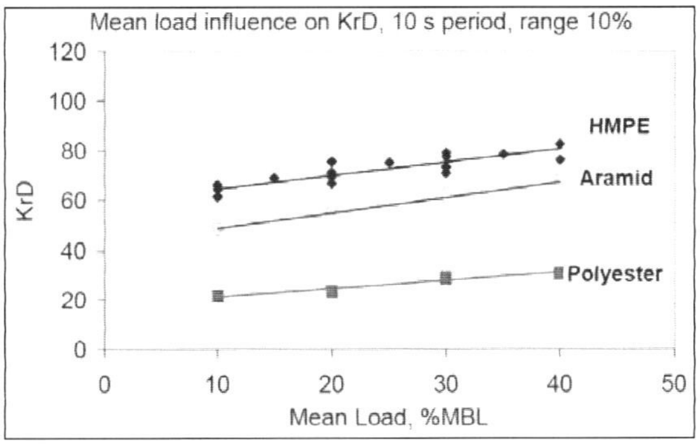

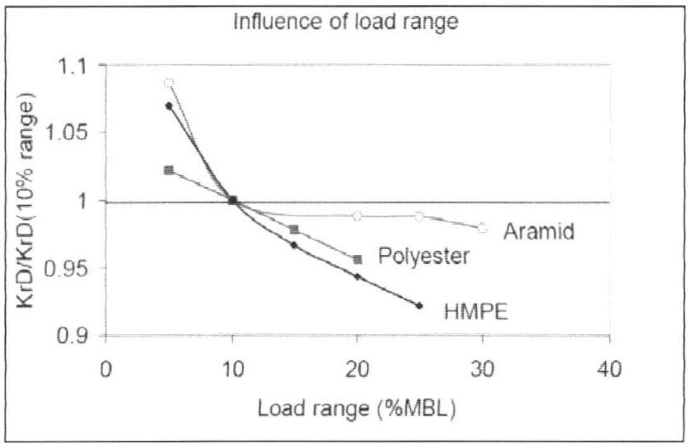

Figure 7. Mean load influence on dynamic stiffness (left); load range influence of dynamic stiffness (right).

repeated for different mean load levels. Mean loads were increased from 10% to 40% and back to 10%. Results see figure 7 (left). As expected the plot shows higher stiffness values for HMPE and aramid. Moreover all materials show increasing dynamic stiffness with increasing mean load.

The load range effect was tested by cycling around 30% mean load and varying the load range from +/-2,5% to +/-15% back to +/-2,5%. The results are shown in figure 7 (right). Here the values are presented as a percentage of the 10% range value. As stated, the graph shows dynamic stiffness decreases with increasing load amplitude.

The previous series of tests were performed at 2 different loading periods, 10s (wave frequency) and either 50s or 100s (low frequency). This allows the influence of the period to be determined directly. Figure 8 shows the dependence of the dynamic stiffness on the loading period, and in particular that HMPE is most sensitive to the loading period parameter.

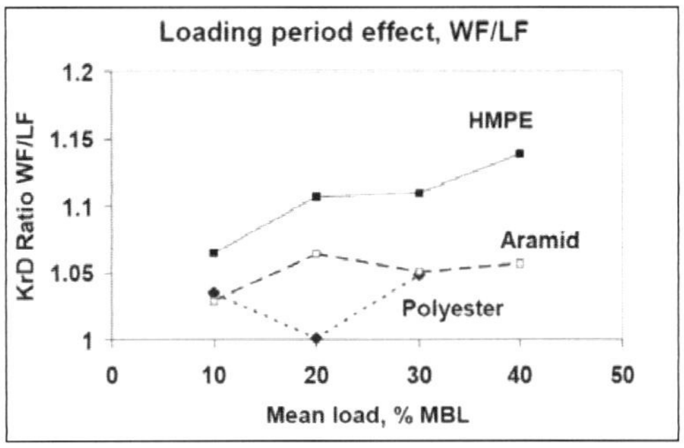

Figure 8. Frequency effect on dynamic stiffness.

The range of amplitudes for a given platform and environment may be much smaller than the tested amplitudes. Also, a platform will never see 100 or more identical loading cycles. This can result in a much smaller effect of load range and frequency on the dynamic stiffness. For polyester, tests have shown that the effects of load range and stiffness can be ignored when the loading sequence is more complex (a superposition of low frequency and wave frequency), and closer to real wave loading than the harmonic (sinusoidal) loading regime used in laboratory testing. The same was found by Francois and Davies[2]. In the table, the dynamic stiffness Kr is expressed in function of the mean load parameter, ignoring load range and frequency.

Material	Dynamic
Polyester	18,5 + 0,33 x (ML%)
Aramid	45 + 0,58 x (ML%)
HMPE	59 + 0,54 x (ML%)

D. CUSTOMIZED ROPE CONSTRUCTION FOR RENEWABLE (WAVE) ENERGY MOORING SYSTEM

Requirements for synthetic mooring tendons

The specific requirements for the mooring ropes in this application are as follows:

- $Kr \leq 10$ during cycling 0-60%MBL
- Minimum Breaking Load 150T - 300T
- 5 year maintenance-free lifetime

Rope design for customized dynamic stiffness

In order to achieve the desired maximum stiffness limit, set by the client, all specifications of the rope were designed from zero. Fiber type, rope construction type and lay length were all optimised for the application.

Suitable fiber materials for ropes with low axial stiffness are today limited to Nylon and polyester. Polyester was found to answer best to the stiffness requirement and was considered sufficiently tested offshore to give high level of confidence, particularly its long term fatigue strength, general resistance to wear and overall reliability.

The most common rope constructions for offshore mooring are ultraline designs where a bundle of parallel sub ropes is combined within a braided cover. This construction is preferred because of the high strength efficiency. The sub rope may be a twisted construction (e.g. 3-strand or 6+1 wire lay construction), or a braided construction such as 8 or 12-strand. A 12-strand construction was chosen for the sub rope because of the higher construction elongation, which assists in a higher elastic response of the rope.

For the 12-strand construction, different twist angles were proposed by TTI using the (FRM) Fibre Rope Modelling program. Next, a series of sub ropes were tested by Bexco to confirm the modelling results, and select the best suitable sub rope design.

Customized rope configuration

A specific rope configuration was requested by the client to improve the efficiency to install the mooring system, and also to take it apart after several years service in a test site.

Bexco has conducted studies into two items that serve to reduce installation time and effort. A special type of Chinese Fingers, that is designed to stay on the rope for the entire lifetime, was strength tested up to 70 Tf, see figure 9 (left). Furthermore, full size pulling eyes have been tested and showed a break strength loss of the main rope of not more than 10-15% . Figure 9 (right). The MBL of the pulling eye itself was found to be 80%-90% of the MBL main rope.

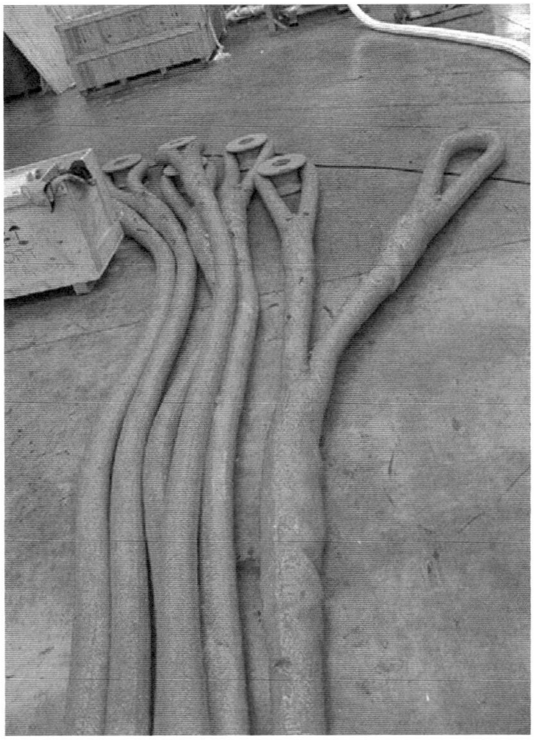

Figure 9. Pictures of Chinese Fingers test (left) and full scale pulling eyes spliced into the rope (right).

E. COMPARISON OF RAW MATERIALS FOR RENEWABLE (WIND) ENERGY MOORING SYSTEM

Requirements for tension members

The specific requirement for the mooring ropes in this application are as follows:

- - 10^6 kN dynamic stiffness
- - <1% elongation (creep + constructional)
- - 25 year fatigue lifetime
- - Minimum Breaking Load 800T

Suitable fiber materials

HMPE, Twaron, Vectran and Carbon are all high strength, high modulus materials that have the potential to replace steel wire in this application. Every material has its own disadvantages that must be overcome, by proper rope design or introduction of new technologies.

It is now widely accepted that many polymer fibers, such as polyester and HMPE, have a longer fatigue life than steel. HMPE has an excellent fatigue resistance, perhaps the best of all fiber materials. In a TCLL (Thousand Cycle Load Level) test Dyneema SK75 achieved a residual strength >100%MBL after 1000 cycles 1-95%MBL. Steel and other polymers like polyester show resp. 60% and 70% res. strength after 1000c 1-80%MBL. An important drawback of HMPE is the creep elongation over time, which is stimulated by increased service load and environment temperature. For this reason static load applications for HMPE may be limited. A less creep sensitive Dyneema type SK78 is available, with the same mechanical properties as SK75. In favour of HMPE is that the material has by now been widely used and accepted for many applications, and that HMPE ropes today can be readily used in the application. Very little public information exist about the fatigue life of aramids (like Twaron), Vectran and carbon ropes. Some sources report a very encouraging potential for aramids, but more testing and gathering of test data is necessary. The lack of fatigue test data has of course to do with the substantial cost associated with these long term tests.

Compared to HMPE, Vectran, aramids and especially carbon fiber show a more brittle behaviour, as a result of their specific chemical formula. This means that in cyclic bending and axial compression, the fatigue lifetime of the fiber is not optimal. Cyclic bending may occur when a tension member with twisted or braided construction is loaded in axial tension-tension fatigue or when a rope is cycled over a sheave. Axial compression fatigue may occur when a rope is cycled to very low tensions (<10%MBL) and some fibers may be forced into compression, while the whole rope is still under tension. In fact, all high modulus fibers (including HMPE) are sensitive to axial compression fatigue. Guidelines for use of high modulus fibers foresee that the minimum tension never drops below 10%MBL. Marine finishes are developed to improve the fatigue life (e.g. for aramids) and also the rope construction has an influence. Another drawback of a brittle material like carbon fiber is a low resistance to yarn-to-yarn abrasion. This is because the transverse properties of the fibers are much less than the properties in the axial direction. In a classic rope construction, internal abrasion may reduce the rope lifetime. For carbon fiber the brittleness is such that, as naked fiber, it can not be used in the classical ropes that are produced today. This would heavily compromise the fatigue life of the rope.

In an attempt to overcome the abovementioned disadvantages of brittle fibers, Bexco is working on composite constructions where the fibers are encased inside a polymer matrix that keeps them from bending and gives overall protection. Many applications using carbon-epoxy structural elements, e.g. the new Airbus A380, have demonstrated that these materials can have an excellent fatigue life.

These composite ropes will have a lot of the desired properties of steel wire, such as low construction elongation, compact design, possibility for a very tough extruded protection cover, etc. Introducing composite materials in the ropes industry will mean a drastic shift away from the classical mooring rope designs towards a mooring tendon that looks like a steel cable, but benefits from the lightweight advantage of polymer materials.

Another challenge lies within the end termination of these composite cables. Naturally they can not be spliced as classical ropes. Bexco is working on an innovative resin socket design for synthetic ropes. The first results show the potential for a superb strength efficiency and fatigue life compared to existing socket designs.

Commercial comparison

This section discusses a commercial comparison between the different proposed fibers, for a specific and recent renewable wind energy project. Only costs of the raw material are compared, production costs are not comprised.

The calculation is made based on the documented fiber stiffness and the findings of P. Davies et al regarding the dynamic stiffness behaviour of HMPE subropes in real wave situation[1]. As Dyneema SK75 (and SK78), Vectran UM and Twaron 2200 all have fiber stiffness in the order of 105 Gpa, the equation *Kr=59+0,54(mean load%)*[1] is used for all materials for the calculation of the rope dynamic stiffness behaviour. This is justified by the results in figure 7 (left) that show the differences in mean load dependency of different (high-modulus as well as medium modulus) materials like HMPE, aramid and polyester is rather small. This approach may not be 100% accurate, but will at least give a close enough approximation.

For a given service load, the minimum stiffness requirement results in a minimum breaking load (MBL) for the rope. The linear weight of the rope is calculated with available efficiencies of Dyneema, Vectran and Twaron ropes recently produced at Bexco. For carbon rope strength and modulus efficiency (yarn - rope) is estimated conservatively at 50%. Finally standard cost info from fiber suppliers is used to make the cost comparison.

Creep data for SK78 provided by DSM is used to calculate the minimum break load necessary to limit the creep elongation to 1%. For SK75 the variation in creep data at these conditions was found to be too high and it was concluded to leave SK75 out of the comparison.

The results of the comparison are shown in figure 10. Note that the cost scale is linear and starts from zero. The results show the important increase for Dyneema SK78 with increasing service load, necessary to limit the creep elongation. Basically the rope MBL needs to be increased to minimise the service load (% of rope MBL) and limit the creep elongation. Materials like Twaron and Vectran slightly benefit from higher service load, but the effect is rather small. Remarkable is carbon fiber rope (based on TohoTenax STS40) that seems to be the most cost efficient in answering to the specific requirements of this application. But this doesn't count for all available carbon fiber types. Because of the very wide range of strength and stiffness properties available, and associated cost, care must be taken when choosing the right carbon type for the application.

The results in this comparison are to a certain extent depending on the rope producer's skills to achieve a good translation efficiency from the single fiber to the full rope. For carbon fiber the efficiency was conservatively taken as 50%, but tests should indicate what can be achieved. Therefore, the foremost important conclusion to take is that, between 5% and 10% service load, which is the service load initially considered for the application, many materials can compete. It is therefore worth to investigate the behaviour and cost efficiency of these materials in practice. In the coming months, sub rope prototypes will be produced and tested for break strength, construction elongation, axial compression fatigue and tension-tension fatigue. Different raw materials will be compared, i.e. Vectran, aramid, HMPE and carbon fiber.

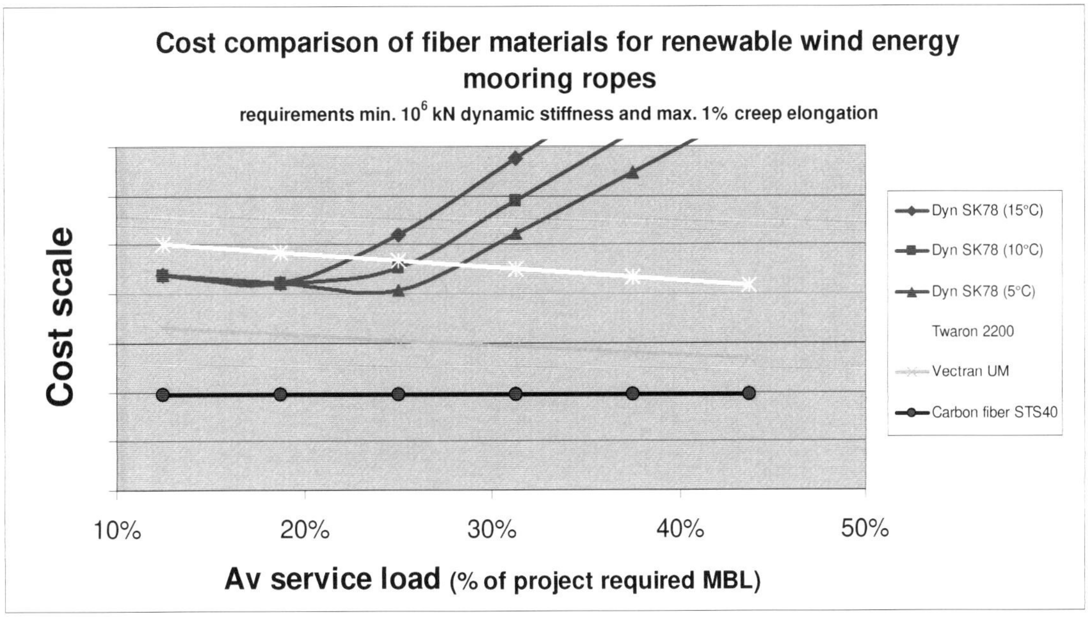

Figure 10. Cost comparison of fiber materials for mooring system for renewable wind energy device.

CONCLUSIONS

In this paper an introduction is given into the rising offshore renewable energy market. It's pointed out that dynamic stiffness is a key requirement for mooring ropes in these systems. Axial stiffness is explained, as well as the difference between static and dynamic load regime. For dynamic stiffness the influencing parameters mean load, load range and frequency are distinguished. It is mentioned that previous research found that the parameters load range and frequency can be ignored in a real wave situation.

A first example case discusses a renewable wave energy device, where the rope construction and rope layout was completely custom designed to fit the application. Studies were done into permanent Chinese Fingers pulling slings and full scale pulling eyes. These special slings assisted in performing the installation most efficiently.

For a specific renewable wind energy device, a comparison is made between different high modulus fiber materials, which material would be the most economically suitable. Different fiber materials like Vectran UM, Twaron 2200, Dyneema SK78 and Carbon fiber (TohoTenax STS40) are compared. Carbon fiber seems to be the most promising for this application, however due to the brittleness this material needs to be part of a composite construction and this still needs to be developed for the ropes industry. Dyneema SK78 shows an important cost increase with increasing service load, but on the other hand Dyneema ropes today have excellent fatigue life and can be readily used in the application. The main conclusion is that in the range of service loads considered, many materials can be competitive, and therefore future work should test the suitability and cost efficiency of these materials on the sub rope level.

ACKNOWLEDGEMENTS

The authors would like to thank Matthijs van Leeuwen of Teijin Aramid, Jac Spijkers of DSM Dyneema, Cheryl Corallo of Kuraray, and Ernst Braches of Toho Tenax for their contributions to this paper.

REFERENCES

[1] Davies et al, (2002), "Synthetic mooring lines for depths to 3000 meters", OTC14246

[2] Francois and Davies, (2008), Characterisation of polyester mooring lines", OMAE2008-57136

(This page intentionally left blank.)

Proceedings of the 16th Offshore Symposium, February 9, 2010, Houston, Texas
Texas Section of the Society of Naval Architects and Marine Engineers

EXPERIMENTAL STUDY ON SUCTION PILE HEAVE RESONANCE DURING LOWERING OPERATIONS

L. Huang, D. Young, J. Zhang, R. Randall
Ocean Engineering Program, Zachry Department of Civil Engineering, Texas A&M University, College Station, Texas

B. Wilde
Intermoor Inc, 900 Threadneedle, Suite 300, Houston, Texas

ABSTRACT

One of the challenges of installing a suction pile is the occurrence of the heave resonance of the pile-cable system during the pile lowering process. To predict and possibly mitigate the resonance of the system, it is crucial to accurately estimate the natural frequency from the total added-mass of the system. The objective of this study is to determine the total added-mass coefficient of a suction pile through model tests.

KEY WORDS: Added-mass coefficient; damping coefficient; heave resonance; suction pile; lowering operation; model test; flow visualization.

INTRODUCTION

Suction piles are a common method of anchoring vessel mooring lines at the seafloor. A suction pile usually consists of a hollow steel cylinder, open at the bottom but closed at the top except for vent valves and a suction port. One of the challenges of installing a suction pile is the occurrence of the heave resonance of the pile-cable system during the pile lowering process. A suction pile is typically lowered to the seafloor on a wire rope cable deployed from a surface vessel's anchor winch or crane. When the heave and/or pitch frequency of the vessel matches the natural frequency of the pile-cable system, the heave resonance of the system may occur, resulting in large oscillations of the pile and thus significantly increasing loads on the lowering cable and device. To predict and possibly mitigate the resonance of the pile-cable system during the lowering process, it is crucial to estimate the natural frequency of the pile-cable system accurately, which in turn needs an accurate estimate the mass of entrapped water inside the pile as well as added-mass water outside the pile. It should be noted that "the mass of entrapped water inside the pile" does not mean the mass of water filling the inside the pile but the mass of water inside the pile moving at the same velocity (or acceleration) as the pile. The objective of this study is to determine the total added-mass coefficient of a pile (including the mass of entrapped water inside and added mass of water outside the pile) through model tests.

The model tests employ an electric motor (actuator) to induce the heave excitation of a pile-cable model in water at certain range of frequencies which cover the resonant frequency of the model. Within the range of frequencies, we determine the resonant (natural) frequency of the pile-cable system depending upon when the model experiences the largest heave. Since the stiffness of the cable is known from the calibration, the natural frequency of the system will reveal the added-mass coefficient of the pile.

EXPERIMENTAL SETUP

Test Tank Facility

The dredge/tow tank in the Haynes Coastal

Engineering Laboratory at Texas A&M University was used for the tests. The tank shown in Fig. 1 is 149.5 ft (45.57 m) long, 12 ft (3.66 m) wide and has a 10 ft (3.05 m) maximum water depth. The sediment pit is an additional 5 ft (1.52 m) deep and is 24.8 ft (7.56 m) long and located about 41 ft (12.5 m) from the weir end of the tank. There is an observation well with windows for observation and video recording. Fig. 2 (left) shows the dredge/tow carriage and the model suction pile actuator that was attached to the front of the dredge/tow carriage.

Actuator and DAQ

The actuator built by Premium Solutions was used for heave excitation of the models while in the tank. As shown in Figure 2 (upper right), it consists of a 220 volt variable speed electric motor, a 20:1 reduction gearbox and the actuating spindle that is attached to the gearbox. The spindle has holes for adjusting the heave stroke to 3 in (7.6 cm), 4 in (10.2 cm), 5 in (12.7 cm), 6 in (15.2 cm), 7 in (17.8 cm), 8 in (20.3 cm), and 9 in (22.9 cm). The actuator assembly was supported by the frame attached to the dredge carriage.

To control the variable speed motor, a voltage regulator was employed. It is a Lenze (AC Tech) SCM Series Sub-Micro drive provided by InterMoor and shown in Figure 2 (lower right). The voltage controller was mounted on a bracket that was situated on the dredge carriage for easy access. The sub-micro drive is equipped with a remote; the drive system has the ability to slowly ramp up the motor.

The cable assembly contains an upper and lower force transducer, upper and lower accelerometer and a spring. The force transducers are 250 lb (113.4 kg) in-line strain gauges and are water submersible. The accelerometers are single directional accelerometers. The first force transducer was placed close to the actuating arm and the second was placed right above the model suction pile cap as shown in Figure 3 (left). This setup was used to determine the different loads in the cable above and below the spring. A spring was inserted in the middle of the cable to simulate the stiffness of an approximately prototype 6,560 ft (2,000 m) cable used for lowering a suction pile to the sea floor so that the model pile may have the equivalent resonant heave period following the Froude number similarity.

For the two accelerometers, the first was placed at the force transducers U-bolt connection and the second at the top of the model as shown in Figure 3 (left). The purpose for the two accelerometers placed at these two locations is to measure the acceleration above the spring and the acceleration of the model suction pile below the spring, which are different as anticipated. In

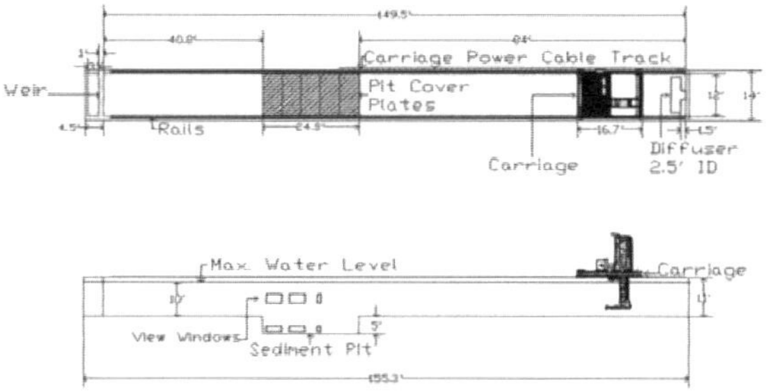

Fig. 1 Elevation and plan view of dredge/tow tank in the Haynes Coastal Engineering Laboratory at Texas A&M University

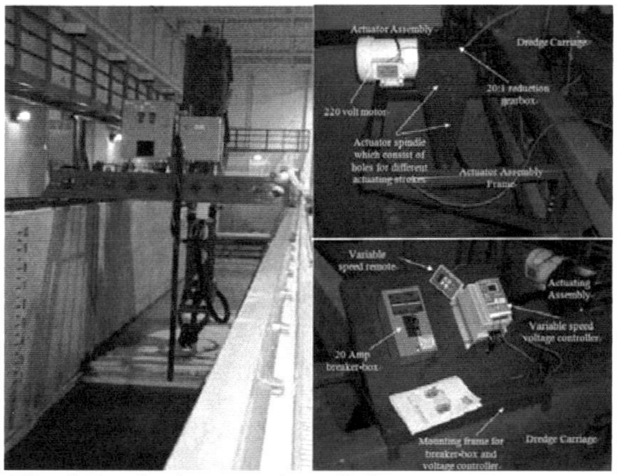

Fig. 2 Dredge/tow carriage (left), setup of actuator (upper right) and variable speed motor controls (lower right)

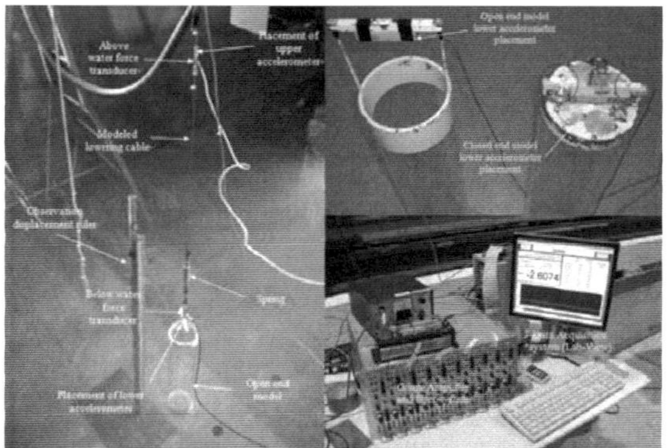

Fig. 3 Model pile with cable assembly (left), accelerometer placement on model suction piles (upper right) and Data acquisition and amplifier setup (lower right)

Figure 3 (upper right), the placement of the lower accelerometer can be seen for the open and close ended models. For the open ended model the accelerometer was attached by using electrical tape on the connection bar for the cable assembly, and for the closed ended model, the accelerometer was attached by using a fabricated acrylic bracket that was attached to the connection for the cable assembly.

Lastly, an observation displacement ruler was put in the tow tank for visually observing the displacement of the model suction pile while it was undergoing oscillation. This visual observation is accomplished by looking through the observation window of the tow tank. The oscillation of the model suction pile was videoed to make sure that the accelerometers were recording correctly.

The force transducers and accelerometers were powered and the output signal amplified to render a good signal. This was done using a strain gauge amplifier shown in Figure 3 (lower right). The strain gauge amplifier is also capable of filtering the high-frequency noise. To record the output signal from the two force transducers and accelerometers, a program called Lab-View was used which recorded the voltage reading into a text file.

Model Parameters

The two suction pile models are shown in Fig. 4 (left) and their basic parameters listed in Table 1. The model scale is 1:32, thus the prototype pile size is 16 ft (4.88 m) in diameter and 96 ft (29.26 m) in length. The main difference between the two models is one with the cap and the other without the cap (open end). There are six pairs of valves with different diameters located on the cap. When all valves are completely sealed, the model is referred as the closed end model. When some or all of these valves are open, the model is referred as partially open end model. For the convenience of description, the valves are numbered from the smallest to the largest in diameter and shown in the Figure 4 (right). The specifications of these valves are summarized in the Table 2. It is noted that a pair of ring reducers can be mounted on the pair of #2 valves to decrease the valve diameter from 1 in (2.5 cm) to 3/4 in (1.9 cm). The latter are named #2' valves. Also modeled was a simulation of a new generation vent valve called the PS valve. The closure plates for this type valve were mounted above the pair of #2' valves as shown in the Figure 3 (upper right). In the standard configuration, the orifices in the pile's top plate are meant to simulate butterfly valves.

Table 1. Basic parameters of two suction pile models

Parameter	Open End Model	Closed End Model
Cylinder length	3 ft (0.91 m)	3 ft (0.91 m)
External diameter	6 in (15.2 cm)	6 in (15.2 cm)
Wall thickness	0.25 in (0.635cm)	0.25 in (0.635cm)
External area	28.274 in^2 (182.413 cm^2)	28.274 in^2 (182.413 cm^2)
Internal area	23.758 in^2 (153.277 cm^2)	23.758 in^2 (153.277 cm^2)
External volume	1017.9 in^3 (16680.4 cm^3)	1017.9 in^3 (16680.4 cm^3)
Internal volume	855.3 in^3 (14015.9 cm^3)	855.3 in^3 (14015.9 cm^3)
Weight of water inside pile	30.9 lb (14.02 kg)	30.9 lb (14.02 kg)
Weight of water displaced by pile plus water inside pile	36.8 lb (16.69 kg)	36.8 lb (16.69 kg)
Dry weight	14.5 lb (6.58 kg)	16 lb (7.26 kg)
Wet weight	8.5 lb (3.86 kg)	9.6 lb (4.35 kg)
Weight of model accessories*	1.06 lb (0.48 kg)	1.06 lb (0.48 kg)
Total dry weight	15.57 lb (7.06 kg)	17.06 lb (7.74 kg)
Gravity center from top	1.29 ft (0.393 m)	1.27 ft (0.387 m)
Buoyancy center from top	1.50 ft (0.457 m)	1.40 ft (0.427 m)

*model accessories include a spring, an underwater force transducer, an underwater accelerometer, a section of cable and connections between the spring and the model.

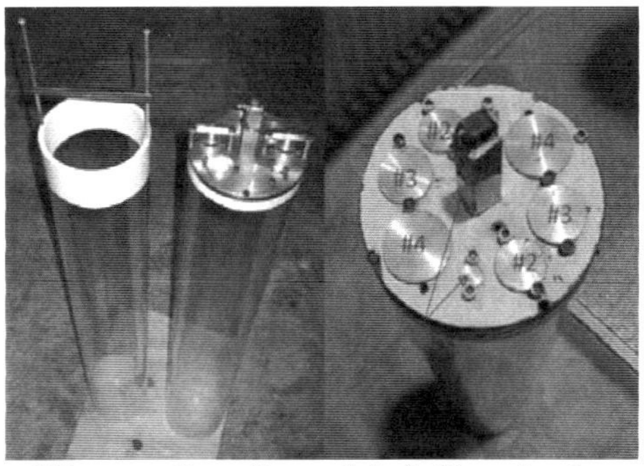

Fig. 4 Two suction pile models (left) and valve pair number for the closed end model (right)

Table 2. Specifications of the valves on the cap

Model Valve Pair No.	Prototype Valve Diameter	Model Valve Diameter	Area of Model Valve Pair
#1	12 in (30.48 cm)	0.375 in (0.953 cm)	0.221 in^2 (1.426cm^2)
#2	32 in (81.28 cm)	1.0 in (2.54 cm)	1.571 in^2 (10.135cm^2)
#2'	24 in (60.96 cm)	0.75 in (1.91 cm)	0.884 in^2 (5.703cm^2)
PS Valve	24 in (60.96 cm)	0.75 in (1.91 cm)	0.884 in^2 (5.703cm^2)
#3	40 in (101.6 cm)	1.25 in (3.18 cm)	2.454 in^2 (15.832cm^2)
#4	48 in (121.92 cm)	1.5 in (3.81 cm)	3.534 in^2 (22.780cm^2)

Dye Injection Procedure

The purpose of the dye injection is to visualize whether or not and how fast the water inside the suction pile model moved with the model while it oscillates. The dye was injected through a ¼ inch plastic hose and a syringe was used to pump the dye from a storage vessel into the plastic hose, as shown in Figure 5. During all tests, dye was released from the end of the hose which was attached to the inside wall of the pile model, as shown in Figure 5. In the case of the closed end model, in addition to the hose attached the inside wall, there was an additional hose also placed inside the model but near the upper cap. The locations of the two hoses in the closed ended model suction pile are shown in Figure 5 (lower right).

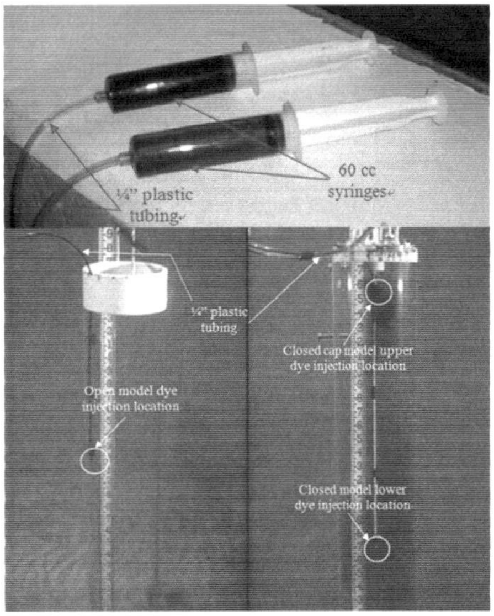

Fig. 5 Device for injecting dye (upper), dye injection layout for open (lower left) and closed (lower right) models

In the case of the open-ended model, a streak of dye from the open end of the plastic hose was observed. Since the open end of the hose was attached the pile model, the dye streak indicates the water inside the pile did not move with the model. However, in the case of the closed end model, no dye streak was observed inside the model, indicating the water moves at the same velocity of the model. In all cases of model testing, the inside of a pile model was always filled with water and no cavitations were observed.

THEORETICAL ANALYSIS

Dynamic System Analysis

To understand the dynamics of its heave, the pile-cable system is simplified and modeled as a single degree of freedom (SDOF) system, which can be described by Eq. (1).

$$M\ddot{X} + C\dot{X} + KX = F(t) \quad (1)$$

where $X, \dot{X}, \ddot{X}$ stand for the heave displacement, velocity and acceleration, respectively, M the total mass of the system (the mass of pile plus the entrained water inside the pile and added mass of water outside the pile), C the equivalent damping coefficient, K the stiffness of the spring, and $F(t)$ the external force. The external force is a function of time and so are the displacement, velocity and acceleration. The overhead dot and dots stand for the derivative with respect to time.

The damped natural frequency ω_d is the real-part solution to the homogenous equation of (1),

$$\omega_d = \omega_n\sqrt{1-\zeta^2} = \sqrt{\frac{K}{M} - \frac{C^2}{4M^2}} \quad (2)$$

where $\omega_n = \sqrt{\frac{K}{M}}$ is the un-damped natural frequency of the system and $\zeta = \frac{C}{2M\omega_n}$ the damping ratio of the system. In general, ζ is very small and less than 0.2 in the case of the open and closed end models, and hence, in these cases the damped natural frequency is very close to the un-damped natural frequency. When the external force is periodic and of an amplitude A_a and frequency ω, the heave amplitude of the pile is equal to

$$A_p = \frac{A_a}{K - M\omega^2 + iC\omega} \quad (3)$$

Eq. (3) shows that when the frequency of the external force is smaller than the un-damped natural frequency the heave of the pile (the underwater acceleration A_2) is in phase of the external force (the above water acceleration defined as A_1). On the other hand, when the frequency of the external force is greater than the un-damped natural frequency, the heave of the pile (the underwater acceleration) is out of phase of the external force (the above water acceleration). When the frequency of the external force is close to the un-damped natural frequency, the phase relation between the heave of the pile and the external force depends on the damping coefficient and the difference between ω_n and ω.

In two cases, the measured above water acceleration (A_1) can be quite different from that induced at the actuator. They are discussed here for better understanding the test results. First, when $\omega > \omega_n$ the external force and heave of the pile are out of phase. Therefore, when the pile is moving up and the cable near the actuator is moving down, the spring is compressed so much that the tension in the spring can reduce to its initial tension. Consequently, the measured acceleration above the spring is smaller than that induced by the actuator. In addition, the tension in the cable below the spring may reach zero, which may hinder the upward motion of the pile and induce the pitch motion of the pile. Secondly, when $\omega \approx \omega_n$, the heave amplitude of the pile becomes significantly greater than the amplitude induced by the actuator. When the pile is moving up and even the actuator is moving up or out of phase by 90 degrees, the spring again is compressed so much that its tension can reach to its initial tension. Consequently, the pile 'pushes up' the spring. If the actuator is almost in phase with the pile, then the measured above water acceleration can be greater than that induced at the actuator. These 'push up' phenomena were observed in the cases of open and closed model near their related resonant frequencies.

Testing Frequency Range

The range of frequencies for testing a model pile is intended to cover its 'unknown' heave resonant frequency. Thus, rough estimate of the resonant frequency of a model pile was made using Eq. (2). Since the total added-mass coefficient (C_a) of each model was not known at this stage, we empirically determined that it was 1.1 for the closed end model and 0.1 for the open end model. The total added-mass coefficient is defined below.

$$C_a = \frac{M_a}{M_0} \tag{4}$$

where M_a is the total added water mass of the system (mass of water displaced by the structure, trapped water mass and added mass) and M_0 is the total mass of the pile (mass of water displaced by the structure plus trapped water mass). The latter was given in Table 1. It is found in calibration procedure of spring that the spring stiffness (K) is 1.5063 lb/in (0.27 kg/cm) and the initial (minimum) tension is 2.222 lb (1 kg). Based on the spring stiffness, the estimated results for both open end and closed end models are given in Table 3. Note that M_1 is the mass of open ended model in air, and M_2 is the mass of closed ended model in air.

Table 3. Theoretical calculation of resonance frequency both in air and in water

	Total mass of the system		f_n	
Open ended model	M_1= (in air)	15.57 lb (7.06 kg)	0.97	Hz
	M_1+0.1M_0= (in water)	19.25 lb (8.31 kg)	0.87	Hz
Closed ended model	M_2= (in air)	17.06 lb (7.74 kg)	0.93	Hz
	M_2+1.1M_0= (in water)	57.54 lb (26.10 kg)	0.51	Hz

It is expected that the resonant frequencies of pile models of a cap with some or all open valves are between the resonant frequencies of the open end model and the closed end one. The range of frequencies for the tests of a model was hence set at about ±0.2 Hz with respect to the related estimated resonant frequency.

Measurement Analysis

The procedures of analyzing measurements of each case of pile model are virtually the same. For an example, the procedures for analyzing the open end model are described below.

After the heave of the open end model reached the steady state, the signals of four sensors (two force transducers and two accelerometers) were recorded for a period of 120 s. Before applying Fast Fourier Transform (FFT) to a time series of measurements, it is multiplied by a unit (height) trapezoidal window function with a ramp of 15 s at both ends. Since the duration of record is 120 s, the frequency resolution is limited to $\Delta f = \frac{1}{T} = \frac{1}{120} \approx 0.0083\text{Hz}$. Fig. 6 shows the measured forces and accelerations in both time and frequency domains for the case of the actuator excitation frequency of 0.89Hz (near resonance).

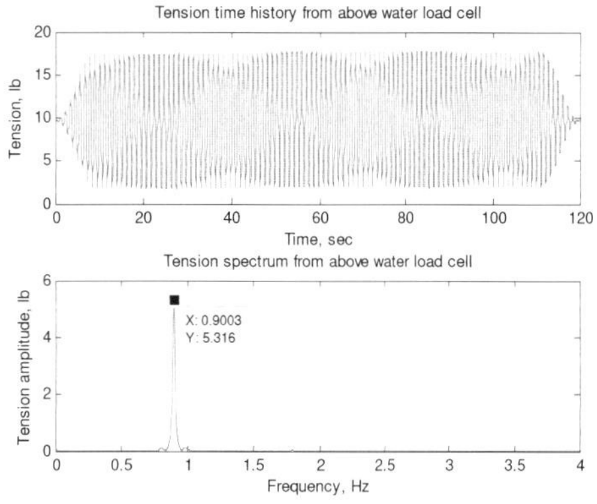

(a) Tension signal from the above water force transducer

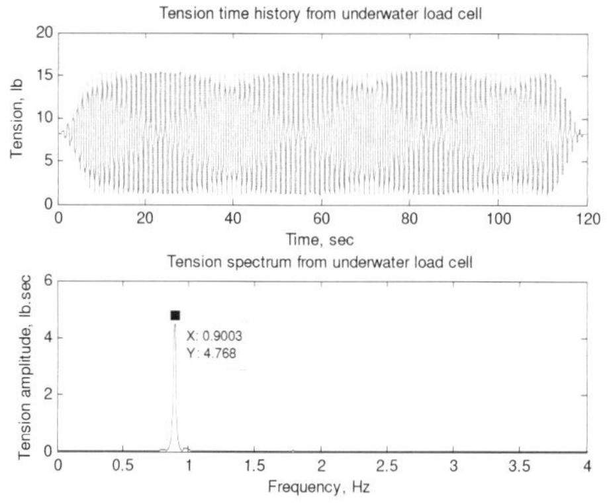

(b) Tension signal from the underwater force transducer

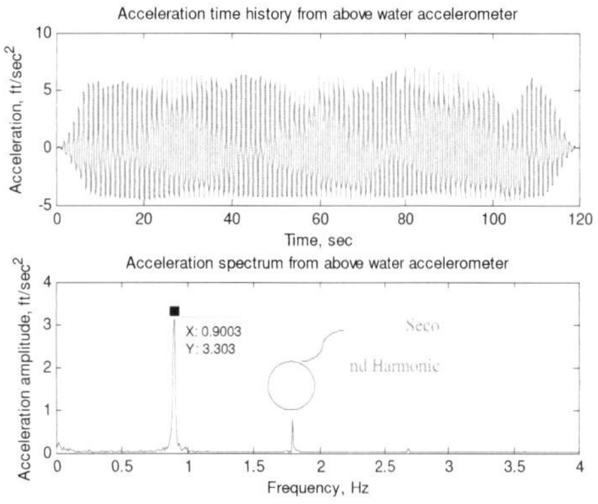

(c) Acceleration signal from the above water accelerometer

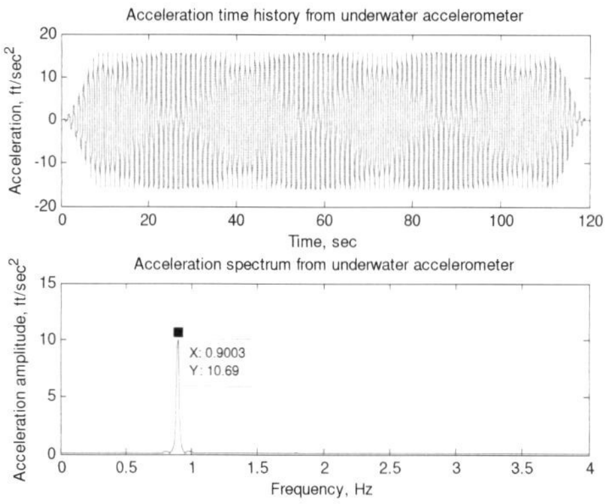

(d) Acceleration signal from the underwater accelerometer

Fig. 6 A 120 second signal record both in time domain and frequency domain at the input frequency f = 0.89 Hz (near resonance)

In these cases, the peak frequency (f_p) for the forces and accelerations are virtually equal to the corresponding excitation frequency of the actuator, which is expected. Considering the energy leaking due to limited duration of the record used in the FFT, the amplitude at the peak frequency is calculated based on 'equivalent' energy concept. That is, the square of the 'equivalent' amplitude is equal to the sum of the square of individual amplitudes nearby the peak frequency. Hence, the equivalent amplitude is greater than that at the peak frequency shown in the figures.

The dynamic tension amplitude above the spring is similar to but slightly greater than the corresponding one below the spring. It is because the tension above the spring also includes the contribution from the spring. However, the acceleration above the spring is significantly smaller than the corresponding one below the spring, especially in the case that the excitation frequency (f = 0.89 Hz) of the actuator is very close to the resonant frequency. This is also expected that the heave amplitude of the model is greatly amplified near its resonant frequency.

It is observed in Figure 6(c) that the measured acceleration by the above water accelerometer is asymmetric (higher positive magnitude and relatively lower negative magnitude) in the time domain. The asymmetry results in a minor peak in the frequency domain, which is located at the frequency twice of that of the major peak and known as the second harmonic response. The asymmetry in the above water acceleration is caused by the 'pushing up' of the pile model. The 'pushing up' occurs when the pile model moves up much more than the cable above the spring and the tension in spring reaches its initial tension (or minimum tension). The 'pushing up' results in greater heave amplitude at the location of the above water accelerometer than that at the actuator. Because of the 'pushing up', the cables above and below the spring can be slack at the moment, which was observed during the tests.

RESULT ANALYSIS

Open End Case

There are slight differences between the frequency of the external force (input frequency) and the measured peak heave frequency of the pile model. The tiny difference is probably caused by the limited frequency resolution in FFT (~0.01Hz) as well as that of actuator controller. The calculated heave amplitudes are obtained by dividing the acceleration amplitudes by the corresponding frequencies. It is shown that the acceleration amplitude at the pile model is much greater than that at the above water accelerometer, especially near the resonant frequency. The calculated heave amplitude above water gradually decreases with the increase in frequency, especially when the excitation frequency is greater than the resonant frequency. However, the calculated heave amplitude very close to the natural frequency can be greater than that at the actuator because of the 'pushing up' by the pile. It is known that the heave amplitude at the actuator remains constant (0.125 ft) and is independent of the excitation frequency. Therefore, the largest calculated heave amplitude at the pile model indicates that the related input frequency is closest to the resonant frequency. The largest heave amplitude in this case is about 4 times of that at the actuator from video record. The calculated heave amplitudes of the open pile model are plotted as a function of peak frequencies in Fig. 7, showing that the resonant frequency is 0.905 Hz.

Closed End Case

Similar to the open end model, the calculated heave amplitude above the spring is smaller than that at the actuator for the excitation frequencies significantly greater than the natural frequency. Near the natural frequency, the calculated heave amplitude above the spring is greater than that at the actuator, which is caused by the 'pushing up' of the pile model. Near the natural frequency, the heave amplitude of the pile model is almost 6 times as that at the actuator from video record. The calculated heave amplitude of the pile model is plotted as a function of the peak

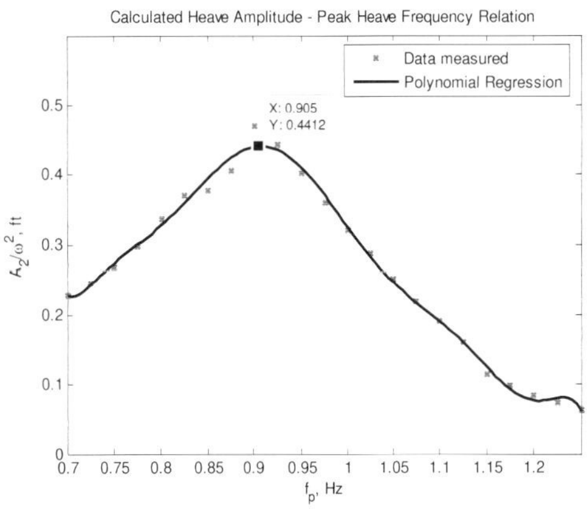

Fig. 7 Relation of calculated heave amplitudes to peak heave frequencies for the open end model tests

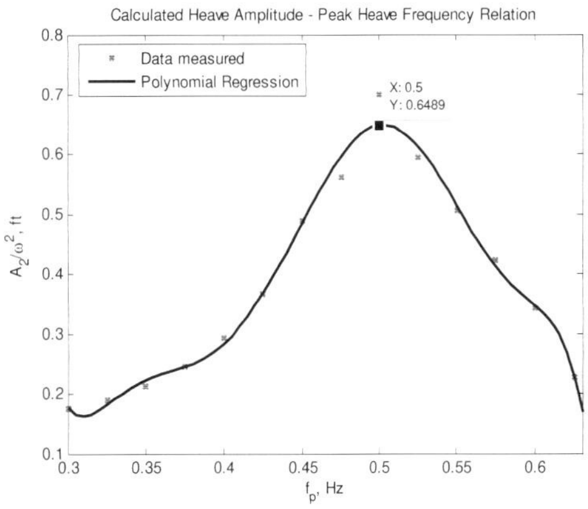

Fig. 8 Relation of calculated heave amplitudes to peak heave frequencies for the closed end model tests

frequency in Fig. 8, showing that the resonant frequency is 0.5 Hz.

Partial Open End Cases

By opening various pairs of orifices (i.e. simulated valves), eight different pile top opening areas plus the special PS valve case could be studied for a total of nine tests. Table 4 summarizes the details of these cases. The most important factor to distinguish these partially open models is the opening area at the cap and presented in the third column. The ratios (γ) of an opening area to the inner cross section area of the pile A = 23.758 in^2 (153.277 cm^2) are given in the fourth column. For brevity, only the results of Case 5 and Case 9 are presented here.

Because of these pile top openings (vent valves), the flow through these openings induces vortices near

Table 4. Nine testing cases of partially open end model conducted in the laboratory

Case No.	Valve Pair Opening	Opening Area	γ
Case 1	#1	0.221 in^2 (1.426 cm^2)	0.9%
Case 2	#2'	0.884 in^2 (5.703 cm^2)	3.7%
Case 3	PS Valve Pair	0.884 in^2 (5.703 cm^2)	3.7%
Case 4	#3	2.454 in^2 (15.832 cm^2)	10.3 %
Case 5	#4	3.534 in^2 (22.780 cm^2)	14.9 %
Case 6	#1, #2, #3	4.246 in^2 (27.393 cm^2)	17.9 %
Case 7	#2, #4	5.105 in^2 (32.935 cm^2)	21.5 %
Case 8	#1, #3, #4	6.210 in^2 (40.064 cm^2)	26.1 %
Case 9	#1, #2, #3, #4	7.780 in^2 (50.193 cm^2)	32.7 %

the cap which significantly dampen the heave of the pile model, especially near their corresponding natural frequencies. The maximum heave amplitudes are about twice of that at the actuator from video records. Because the heave amplitudes of the pile model in these two cases are small in comparison with those in the previous two cases, the 'pushing up' from the pile model to the spring no longer occurs. Consequently, the heave amplitude above the spring monotonically decreases with the increase in the excitation frequency. During the tests of these two cases, the phenomena of 'slack cable' and the pitch of the models were not observed when the excitation frequency was close to the related natural frequency. The heave amplitudes of these two cases are plotted as a function of peak frequencies in Fig. 9 and Fig. 10, showing that the resonant frequencies are 0.555 Hz and 0.71 Hz respectively.

Summary of All Cases

The open and closed end models can be viewed as special cases among the partially open cases. That is, $\gamma=1$ for the open end model and $\gamma=0$ for the closed end model. All the results of the eleven cases are summarized in Table 5. The resonance frequency (f_{res}) and the resonance circular frequency (ω_{res}) for partially open end cases are determined by the same way as illustrated in open and closed end cases. The total weight of the system (W'), is calculated from the resonance frequency using the following relationship:

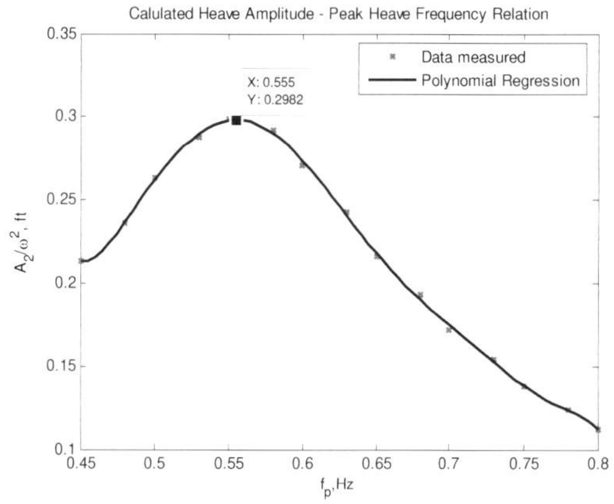

Fig. 9 Relation of calculated heave amplitudes to peak heave frequencies for the partially open end model tests (Case 5)

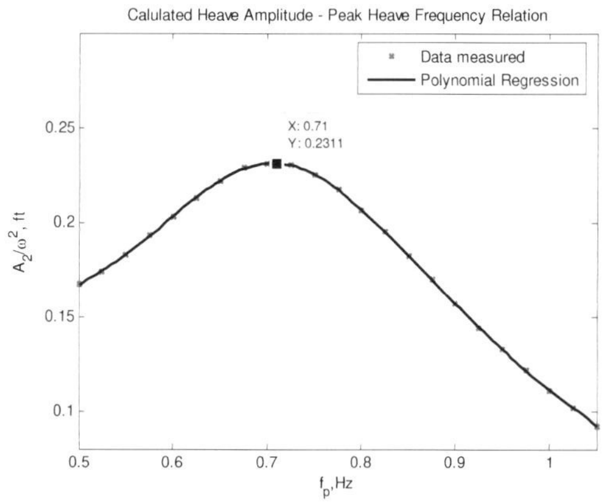

Fig. 10 Relation of calculated heave amplitudes to peak heave frequencies for the partially open end model tests (Case 9)

$$W' = \frac{K}{\omega_{res}^2} g \qquad (5)$$

The weight of model pile (W_{pile}), weight of water inside model pile (W_{in}), and weight of water displaced by pile plus interior entrapped water (W_0) are identical except the open end model, shown in Table 1. The total added water weight ($W_a=W'-W_{pile}$), is given in Column 4. The exterior added water weight ($W_{a,ex}=W'-W_{pile}-W_{in}$), and exterior added-mass coefficient ($C_{a,ex}=W_{a,ex}/W_0$) are given in Column 5 and Column 6. Assume $C_{a,ex}$ are constant for closed end model and partially open end models. The interior entrapped water weight ($W_E= W'-W_{pile}-W_{a,ex}$), and entrainment ($E= W_{a,in}/W_{in}$) are given in Column 7 and Column 8. If E is less than zero, let E is equal to zero and recalculate the exterior added mass coefficient.

Table 5. Summary of results of all the model test cases

γ	f_{res} (Hz)	W'	W_a	$W_{a,ex}$	$C_{a,ex}$	W_E	E	TR	Note
0.000	0.5	58.91 lb (26.72 kg)	41.85 lb (18.98 kg)	10.95 lb (4.97 kg)	0.298	30.90 lb (14.02 kg)	1.000	5.4	Pile top closed
0.009	0.505	57.75 lb (26.19 kg)	40.69 lb (18.46 kg)	10.95 lb (4.97 kg)	0.298	29.74 lb (13.49 kg)	0.962	5	
0.037	0.51	56.62 lb (25.62 kg)	39.56 lb (17.94 kg)	10.95 lb (4.97 kg)	0.298	28.61 lb (12.98 kg)	0.926	4.8	
0.103	0.54	50.51 lb (22.91 kg)	33.45 lb (15.17 kg)	10.95 lb (4.97 kg)	0.298	22.50 lb (10.21 kg)	0.728	4.2	
0.149	0.555	47.81 lb (21.69 kg)	30.75 lb (13.95 kg)	10.95 lb (4.97 kg)	0.298	19.80 lb (8.98 kg)	0.641	3.2	
0.179	0.585	43.03 lb (19.52 kg)	25.97 lb (11.78 kg)	10.95 lb (4.97 kg)	0.298	15.03 lb (6.82 kg)	0.486	3	
0.215	0.61	39.58 lb (17.95 kg)	22.52 lb (10.21 kg)	10.95 lb (4.97 kg)	0.298	11.57 lb (5.25 kg)	0.374	2.9	
0.261	0.64	35.96 lb (16.31 kg)	18.90 lb (8.57 kg)	10.95 lb (4.97 kg)	0.298	7.95 lb (3.61 kg)	0.257	2.85	
0.327	0.71	29.22 lb (13.25 kg)	12.16 lb (5.52 kg)	10.95 lb (4.97 kg)	0.298	1.21 lb (0.55 kg)	0.039	2.8	All valves open
1.000	0.905	17.98 lb (8.16 kg)	2.41 lb (1.09 kg)	2.41 lb (4.97 kg)	0.066	0.00 lb (0.00 kg)	0.000	6.4	Fully open
0.037	0.505	57.75 lb (26.19 kg)	40.69 lb (18.46 kg)	10.95 lb (4.97 kg)	0.298	29.74 lb (13.49 kg)	0.962	5	PS valves used

The 9th column lists the Transmissibility (TR) which is defined as the ratio of the heave amplitude of pile at resonance observed from video records to the heave amplitude at the actuator (1.5 inches (3.81 cm) heave amplitude throughout all the tests). The last row lists the results of the case of only the PS valve opening. Although the opening PS valve ratio ($\gamma = 0.037$) is greater than that of the case of opening Valve #1 ($\gamma = 0.009$), their natural frequencies and the added-mass coefficient are identical. This is because the PS valves have closure plates hanging about one inch above the orifice. The closure plates hinder the flow through the orifice and effectively reduce the opening area ratio. In general, the larger the opening area ratio is, the smaller the added-mass coefficient is, and consequently, the larger the natural (resonant) frequency is. The related results are plotted in Figs. 11, 12 and 13, respectively. The trend observed in these two figures is as expected. That is, a larger opening ratio results in less mass of water moving with the same velocity as the pile model.

CONCLUSIONS

Based on the results of model tests, several preliminary conclusions were derived and are given below.

1) Due to the presence of the openings at the cap, part of the water inside the pile model flows through the openings and hence does not move at the same velocity and acceleration with the pile. The larger the opening ratio is, more flow passes through the openings or less mass of water is entrapped (or moves with pile). Hence, the total mass of a model reduces with the increase of the opening ratio. Consequently, the natural frequency of a model increases with the increase of the opening ratio. In the case of the closed end model, the water inside the pile is 100% entrapped within the pile. Hence, its natural frequency is the lowest.

The entrained water ratio varied from near 100% for the closed pile top case, to 96% with a pair of 24 in (61 cm) (prototype size) valves opened, 65% with a pair of 48 in (122 cm) valves opened and 4% when four pairs of [12 in (30.5 cm), 32 in (81.3 cm), 40 in (101.6 cm) and 48in (122 cm)] valves were opened. However, this calculation of the mass of entrapped water is based on the assumption that the outside added mass coefficient ($C_{a,ex}$) remains constant when the open area ratio changes. As we know, the flow pattern around the upper end of the pile may change due to the presence of opening valves. It is likely that the outside added mass coefficient may decrease with the increase in the opening area ratio.

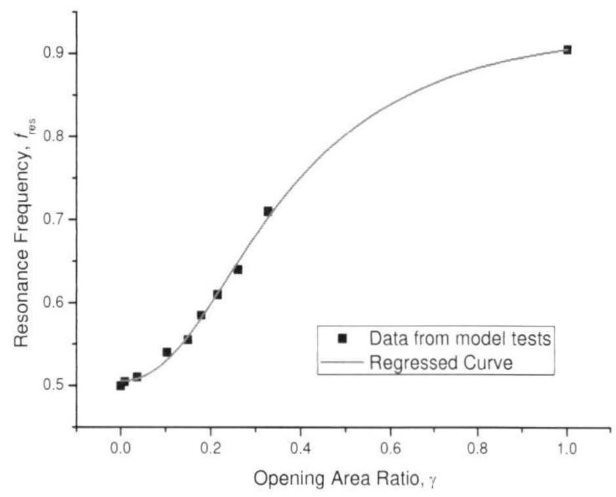

Fig. 11 Relation of opening area ratio to resonance frequency

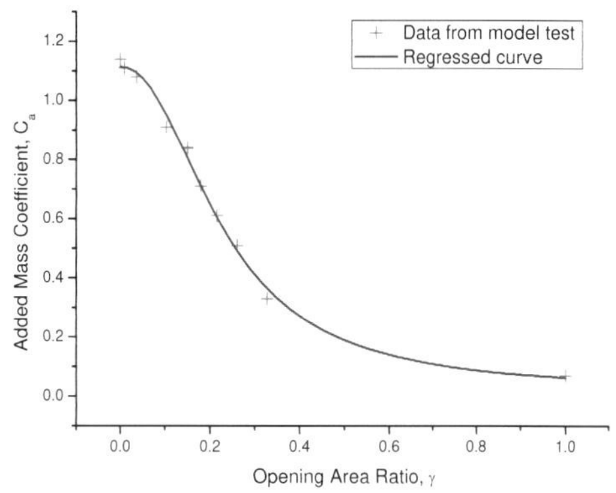

Fig. 12 Relation of opening area ratio to total added mass coefficient

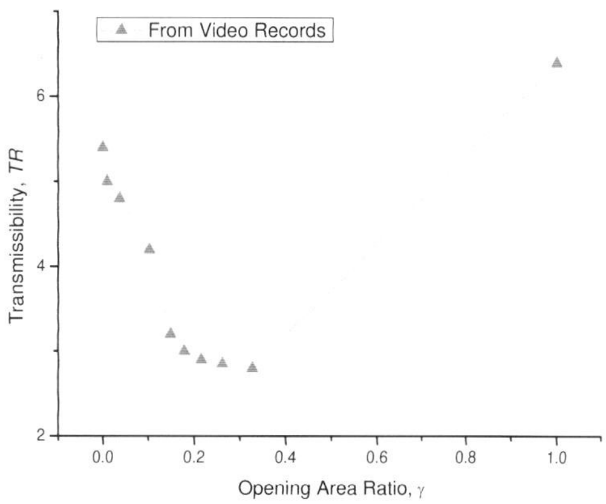

Fig. 13 Relation of opening area ratio to transmissibility

2) Due to the presence of the openings at the cap of the pile, water flows through them when the pile is oscillating. This flow generates vortices above and below the cap and hence results in additional resistances to the heave of the pile. Consequently, the resonance heave amplitudes for the partially open models are significantly smaller than that of either open or closed model. However, in order to significantly reduce the resonant heave amplitude the opening ratio has to be greater than 0.15.

For the closed pile top case, the critical damping ratio was found to be 9.4%. This increased to 16.5% when both 48 in (122 cm) (prototype size) valves were opened. With the pairs of 12 in (30.5 cm), 32 in (81.3 cm), 40 in (101.6 cm) and 48 in (122 cm) opened, the critical damping ratio peaked at 19.1%.

3) The greatest resonant heave amplitude was observed in the tests of open and closed models when the excitation frequency is closed to the related natural frequency. The large heave amplitude results in the 'push up' phenomenon when the pile model pushes up the spring and cable. The 'push up' by the pile may significantly reduce the tension in spring and cable. The reaction force from the spring and cable may hinder the upward motion of the pile model and hence may induce pitching of the pile.

4) Increased damping that reduces resonant response would be viewed as a desirable feature when the ratio of the system natural period to the peak excitation period called beta ($\beta = T_n/T_p$) is less than the square root of 2. When lowering relative low mass objects such as well control Christmas trees, this would be the case.

ACKNOWLEDGEMENTS

This study was sponsored by the InterMoor Inc. and conducted at the Haynes Coastal Engineering Laboratory at Texas A&M University. We gratefully acknowledge the approval of publication of this study by the InterMoor Inc. We also thank the Premium Solutions Inc for building the actuator and testing models, and laboratory technician John Reed and graduate student John Henrikson for technical help of electronic equipments.

REFERENCES

[1] CHOPRA, A. *Dynamics of Structures.* New Jersey: Prentice-Hall, 2001.

Proceedings of the 16th Offshore Symposium, February 9, 2010, Houston, Texas
Texas Section of the Society of Naval Architects and Marine Engineers

NEW METHOD FOR TRANSPORT & INSTALLATION OF HEAVY SUCTION ANCHOR OR STRUCTURES IN EXTREME DEEP WATERS >2000m

K.J.Vermeulen, R.L.Krabbendam
Jumbo Offshore Vof, Rotterdam, The Netherlands

ABSTRACT

This paper gives an insight in the evolution of deep water lifting capabilities on the J-class installation vessels of Jumbo. Two existing methods that are used for recently executed deepwater installation projects are described together with the pro's and cons of each system. Based on this comparison, a new concept for installation of heavy structures in ultra deep water is introduced. In order to check the feasibility of this new concept, an in-house analysis campaign has been executed. The first results and discussion of these results are presented at the end of this paper.

INTRODUCTION

Driven by the requirements of the offshore installation industry to install larger and heavier structures in greater water depths, Jumbo Offshore has successfully installed and used a Deep Water Deployment System on one of their J-class Heavy Lift vessels. The system consists of a traction winch and a storage winch that are positioned in the cargo hold. The winch wire is fed through the cranes via a letterbox in the cargo hold. For each of the two 900t mast cranes such a system is installed. In this way, cranes can be used directly for pick-up, over boarding and installing structures on the seabed. By means of this system heavy loads can be installed on the seabed in water depths up to 3000m. However, due to the relative large weight of steel wire, a large part of the deep water lifting and lowering capability is consumed by the submerged weight of the steel wire at larger water depths. To reduce this self-weight consumption and increase the lifting capabilities to more or less unlimited water depths, a deepwater installation method using pennants for lowering of structures to the seabed has been developed.

EXISTING DEEP WATER CAPABILITIES

The size and crane capabilities of the Jumbo J-Class vessels offer a large flexibility for executing sub sea installation work.

Since 2003, Jumbo Offshore has used two methods for installing heavy sub sea structures each with their pro's and cons.

Wet handshake method

When the crane winchdrums do not have sufficient wire carrying capacity to lower the hookblock to larger water depth, one can use the wet handshake method. By means of this method, the sub sea equipment is lifted from the deck and through the splash zone with one or both 900 tons mastcranes and lowered to about 100 m water depth. At 100 m water depth, the load is transferred from the crane to a deep water lowering drum winch on the side of the vessel. The picture below shows a drum winch located on Port side of the vessel. This drum winch has a capacity of 1500 m wire length with a diameter of 90 mm and an SWL of 180 t (FOS 3).

Fig. 1 large drum winch installed at PS of Jumbo vessel

This method has been successfully applied for the pile installation work on the Kikeh project described in reference [1]. The disadvantages of this method are:

- Complexity: the wet handshake at 100 m water depth requires extra rigging gear as well as the use of a Work Class ROV and is therefore a delay point in the total installation sequence.
- Limited capacity: when using a drum winch, the capacity in terms of wire length as well as in terms of Safe Working Load (SWL) are limited. Max. wire diam. is presently limited to approx. 140 mm and 2500 m length because of manufacturing capabilities.

When using a drum winch, the wire length limitation originates from the fact that the number of layers that can be spooled safely on the drum are limited to approx. 6-8 layers, hereby limiting the max. wire length, unless the drum diam. and winch size is increased significantly.

The SWL limitation originates from the fact that the submerged weight of the steel wire consumes more than 40% of the SWL at depths beyond 2000 m

Own cranes

To avoid the use of a wet handshake for installing sub sea equipment on the seabed, the wire carrying capacity of the crane winches had to be increased. By doing this, the equipment can be picked up from the hold or deck by means of the 900 tons mastcranes and lowered directly to the seabed. However, by using a conventional drum winch, the amount of wire to be spooled is still limited to the maximum allowable number of layers on the drum winch. This would result in excessive drum dimensions for large wire lengths. Therefore, Jumbo Offshore has decided to use a traction winch in combination with a storage winch drum to increase the wire carrying capacity of both cranes. The advantage of a traction winch storage winch drum combination is given by the fact that the required line pull of 70 t is generated by the traction winch and reduced to a continuos linepull at the storage winch drum. The wire will be spooled at a constant tension of about 5 tons on the storage winch drum. In this way more than 14000 m of 50 mm steel wire can be safely spooled on the storage winch drum. There is a manufacturing limitation of approx. 240 t of steel wire rope weight in one length by the manufacturers.

In the August of 2009 the first two traction winch-storage drum combinations have been installed on the Heavy Lift Vessel Fairplayer. The storage drum and traction winch are shown in the picture below.

Fig. 2 Storage winch drum and traction winch

The system has been successfully deployed during the installation of the Free Standing Hybrid Risers on the Cascade and Chinook field (Gulf of Mexico) in December 2009 in 2500 m water depth

NEW METHOD

Having solved the wire carrying capacity limitation, the submerged weight of the steel crane wire is still the limiting factor in terms of extreme deepwater weight carrying capacity. This problem can be solved by means of the new method of using pennants for lowering heavy structures to larger water depths. When using the installation of suction anchors as an example, the polyester or mooring lines constructed of any other type of fibre, can be used as pennant. Due to the fact that the fibre ropes are neutral buoyant, the self-weight disadvantage is resolved. The submerged tackle weight is now limited to the lowering depth of the hoisting tackle, beyond that we do not have the submerged weight disadvantage anymore. In the pictures below, the installation steps are indicated.

Step 1: hooking on and picking up suction pile from the lower hold.

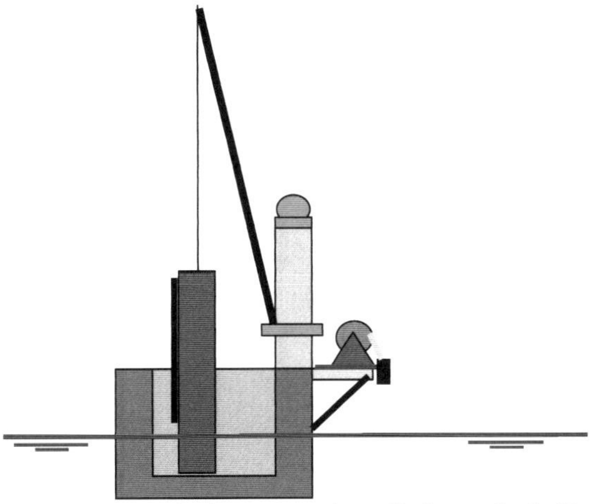

Fig.3 step 1: picking up suction pile from the hold

Step 2: slewing the suction pile adjacent to the hang off structure and connecting the first polymer line to top end of the anchor chain of the suction pile using a connector.

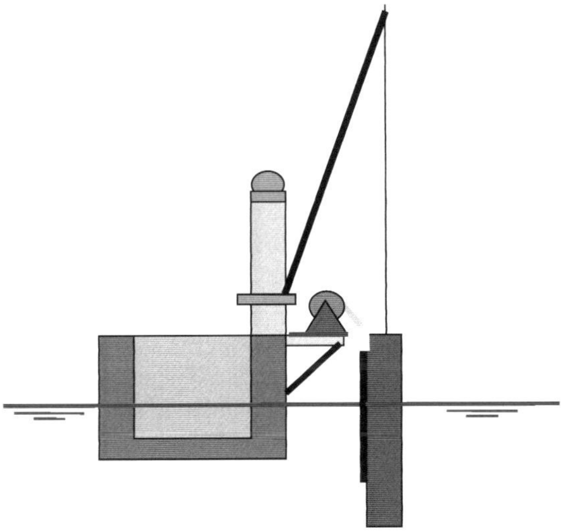

Fig. 4 step 2: connecting mooring rope to mooring chain

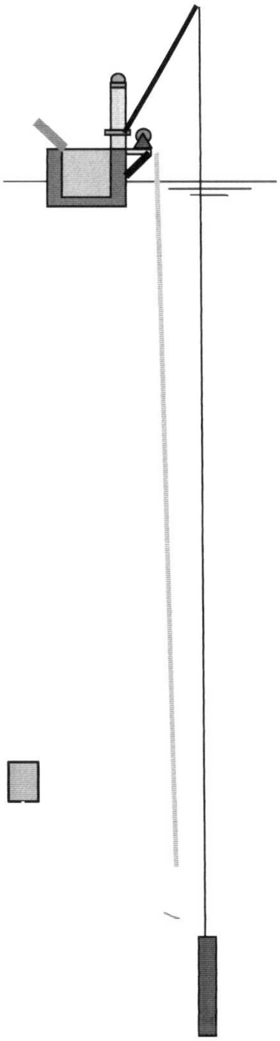

Fig. 5 step 4: transfer suction pile weight to pennant

Step 3:

1. Transferring the suction pile from the crane to the first poly mooring line supported by the hang off structure.
2. Releasing the crane main block from the suction pile and suspending the suction pile from the first polymer line connected to the hang off structure while retrieving the crane main block to the deck.

Step 4:

1. Connecting a second polymer mooring line to the first polymer line at the deck level. Lifting the first polymer line with the second polymer line connected using the crane and lifting both polymer lines out of the hang off structure.
2. Deploying the second polymer line using the reel drive and lowering the suction pile supported by the first polymer line to a second depth using a

deepwater deployment system of the crane and connecting the end of the second polymer line to the hang off structure.

3. Transfer load of the first polymer line with suction anchor to the second polymer line connected to the hang off structure and retrieving the crane main block to deck; This procedure could continue with 3rd, 4th and more polymer mooring lines (depending on water depth and lengths of polymer mooring lines.)

DYNAMIC ANALYSIS

Lift in air

Depending on the nature of the lifts, two boundary conditions can be identified during picking up and overboarding the suction piles.

1)Lifting the suction pile restrained with respect to the vessel.

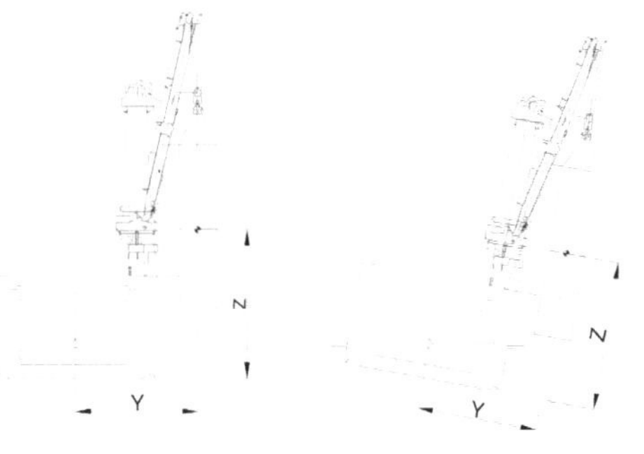

Fig. 6 restrained suction pile

2)Lifting the suction pile unrestrained with respect to the vessel.

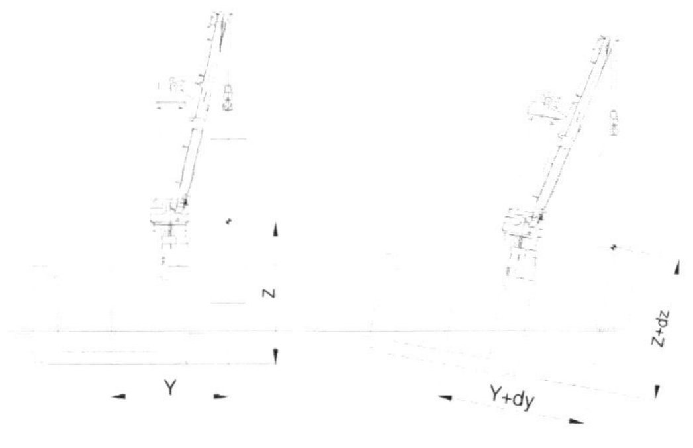

Fig. 7 free suction pile

The behaviour of the vessel and suction pile will be different in both situations. In the first situation there will be no pendulum motion of the lifted suction pile. The mass of the suction pile remains a rigid part of the total vessel mass during picking up and overboarding. In the second situation, the suction pile is free to swing once it is lifted from the tanktop. In terms of vessel behaviour, both situations have a different effect on lifting stability and dynamic behaviour.

lifting stability

In the first situation the suction pile does not influence the stability of the vessel because the position with respect to the vessel is fixed when the vessel heels. In the second situation, the suction pile does influence the stability of the vessel. When the vessel heels, the transverse distance of the suction pile to the center of gravity of the vessel increases (Y+dy) due to the pendulum motion of the suction pile. As a result, the suction pile introduces an additional heeling moment when heeled. This effect can be translated into a virtual loss of initial stability (G'Mt) by assuming that the mass of the suction pile is located at the crane tip position. For the assessment of lifting stability the 2nd situation is the governing case.

dynamic behavior

In the first situation the dynamic behaviour of both suction pile and vessel can be approached by a single body analysis. The second situation should be approached by a coupled 2 body analysis. The swinging suction pile will have its influence on the Roll motions of the vessel and vice versa, the Roll motion of the vessel will influence the suction pile motion.

In reality, the situation will be in between; in order to prevent excessive pendulum motions, the suction pile will be restrained to the vessel by means of an extensive tugger arrangement. But there always will be some relative motion between the hook load and the vessel. See figure 8 below for a 'tuggered' load.

Fig. 8 restrained hookload

The limited restrained hookload motions together with the limited Roll and Pitch motion criteria for Picking up and over boarding the suction pile both result in neglect-able influence of the suction pile motions on the vessel behaviour. Therefore, the actual the dynamic behaviour of vessel and 'tuggered' suction pile can be approached by single body motion analysis. The tugger arrangement should be dimensioned based on a 2 body analysis of vessel and suction pile restrained by modelled tugger wires and winch behaviour.

subsea lift

From an operational point of view, the new method of using fibre rope pennants in combination with the deep- water deployment system reduces the complexity (no wet handshake) and increases the carrying capacity at depth. From a technical point of view, following issues were investigated.

1. the effect of the changed properties of the hoisting system on the sensitivity for certain excitation periods; natural period analysis
2. the effect of this sensitivity on the dynamic behaviour; dynamic amplification analysis

natural periods

In addition to the reduced weight/volume ratio, the stiffness properties of the pennants will affect the natural period of the hoisting system including pennants. The natural period of the hoisting system is an important parameter, since this parameter determines the dynamic amplification of the system under harmonic excitation in waves. To get an insight in the order of magnitude of the natural period as a function of pennant length, pennant material and lowering depth, the natural periods are calculated by means of the formulation fore axial natural period of a spring-mass system:

$$T_0 = 2\pi\sqrt{\frac{M}{K}}$$

in which:

T_0 natural period [s]

M mass of system [t]

K stiffness of hoisting system [kN/m]

The mass of the system includes:

1. dynamic (added + entrapped) mass
2. hoisting wire submerged mass
3. structural mass of hook load

The total stiffness of the hoisting system is given by:

$$\frac{1}{K} = \frac{1}{K_c} + \frac{L_{HW}}{EA_{HW}} + \frac{L_P}{EA_P}$$

in which:

K_c vertical crane stiffness [kN/m]

L_{HW} length of hoisting wire [m]

L_P length of pennant [m]

E modulus of elasticity [N/mm^2]

A cross sectional area of cable [mm^2]

To get an insight in the effect of the pennant material and lowering depth, the natural period analysis is conducted for 3 pennant materials. The load for the analysis is a 300t suction pile with a diameter of 6m, a length of 27m, an entrapped mass of 756t and added mass of 109t.

Configuration	EA tackle [mN]	EA pennant [mN]
10 falls 1000m lowering	1260	
10 falls 1000m + 1000m aramide pennant		735
10 falls 1000m + 2000m aramide pennant		
10 falls 1000m + 1000m polyester pennant		235
10 falls 1000m + 2000m polyester pennant		
10 falls 1000m + 1000m steel pennant		582
10 falls 1000m + 2000m steel pennant		

Table 1 analysis cases

From the table, it can be observed that the stiffness of the Aramid pennant is more than 3 times larger with respect to polyester. It should be noted that, as a result of the parallel strand construction (FibreMax) the stiffness of the Aramid pennant is mainly material stiffness, while for the braided polyester and steel pennant, the total stiffness is the sum of construction and material stretch.

In the figure below, the natural periods of the hoisting system are presented on the Y-axis as a function of lowering depth on the X-axis.

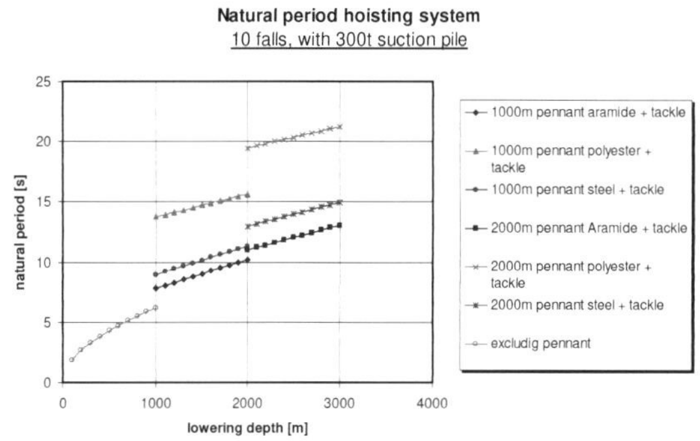

Fig 9 natural periods

From the figure, it can be observed that the natural periods of the Aramid and steel pennant combinations are in the same order of magnitude. The natural periods of the polyester pennant-tackle combinations are above 14s for the complete depth range from 1000 m to 3000 m. The figure clearly indicates the shift in natural period when the load is transferred from the relatively stiff hoisting tackle to the soft pennant at 1000 m and 2000 m lowering depth.

dynamic amplification

To investigate the effect of the crane tip motions on the load pennant loads, a simulation model of the vessel-crane-hoisting tackle-pennant-suction pile combination has been constructed in the time domain analysis program Orcaflex.

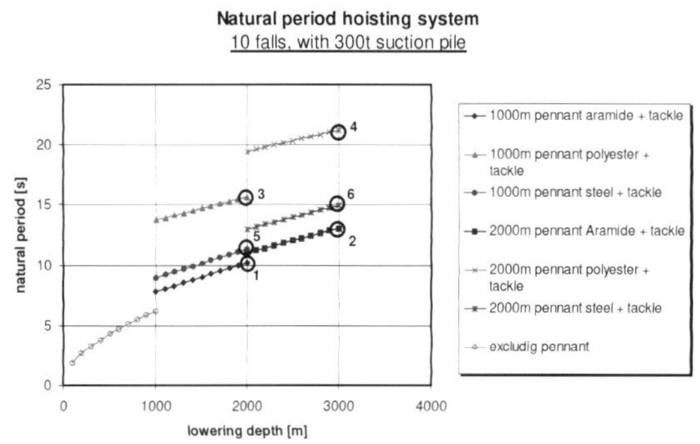

Fig 10 selected cases for dynamic analysis

Within Orcaflex the vessels' Response Amplitude Operators are input, the crane wire and pennant are modeled as a number of elements each with their own mass and stiffness. The suction pile is modeled as an element with mass, added mass, entrapped mass, drag and volume.

For the Orcaflex analysis, 6 cases are selected from the natural period analysis. The selected cases are presented in the figure below.

The Orcaflex simulations are performed in regular waves with a height of 4 m and a period of 11 s coming in from slightly (15 deg) off-bow direction. The reasons for using regular waves for the analysis are twofold:

1. Computation efficiency; due to the fact that statistical representation of extremes in waves are determined outside the calculation model by selecting the appropriate wave height for the regular wave train, the length of the simulations can be limited to the time needed to converge to a numerical stable behavior.
2. Exact tuning on natural period; when using regular waves with a period that matches exactly the natural period, the resonance behavior can be addressed.

Furthermore, Swell types of waves do approach the regular wave, due to the dispersive effect of the waves.

The results of the Orcaflex analysis are presented in the figure below. The results are presented in terms of percentage dynamic amplification of the force in the pennant with respect to the mean force at the top of the pennant.

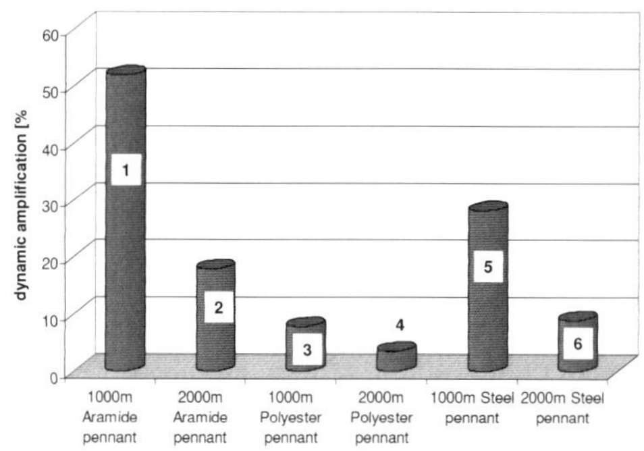

Fig. 11 Dynamic amplification

From the figure, it can be observed that the 1000 m Steel or Aramid pennant in combination with 1000 m hoisting tackle induces the largest dynamic amplification. This is the result of crane tip motions with periods (11s) near the resonance period of the hoisting system. The softer combinations with Polyester pennants show all relative small dynamic amplification values.

DISCUSSION

From figure 7 it can be concluded that, when the hoisting system of pennant and hoisting tackle is excited with crane tip motions having periods around the natural period of the hoisting system, unacceptable dynamic amplification occurs with respect to the mean pennant force. Hence, crane tip motions with a period near the resonance period of the hoisting system should be avoided.

Given the wave conditions and the ship motion behaviour, the question that arises is: how can we avoid resonance behaviour of the system in the case of Aramide of Steel pennant combination?

To answer this question, we have to take a closer look on the analysis results of the polyester pennant combination.

From the natural period analysis of the Polyester pennant combination, it can be concluded that the softer pennant system has natural periods above 14s for all combinations. As a result, the dynamic amplification for these cases is relatively low (below 10%) as well as indicated in figure 7.

Based on this observation, it can be concluded that, if we are able to shift the natural period of the stiffer hoisting system out of the range of the crane tip motions, the dynamic amplification will decrease to acceptable values.

This natural period shift can be achieved by adding a soft spring to the hoisting system. As demonstrated by means of the formulation for the total combined stiffness, the combined stiffness of the hoisting system will be governed by this soft spring. It should be noted that this is true under the condition that the relative difference between the stiffness of the hoisting system and the stiffness of this added soft spring is significant.

In practice, this soft spring is known as a passive heave compensator. The passive heave compensator consists of nitrogen filled cylinder and a plunger.

When added between the load and the crane hook or hang off frame, this passive heave compensator act as a spring-damper combination.

To demonstrate the effect of this passive heave compensator on the natural period of the hoisting system with the 1000m Aramide pennant, the natural periods are calculated for this combination including passive heave compensator.

The selected passive heave compensator has a Safe Working load of 430 t and is adjusted in such a way that the half stroke value corresponds with the submerged weight of the suction pile. The effect on natural period is presented in the figure below.

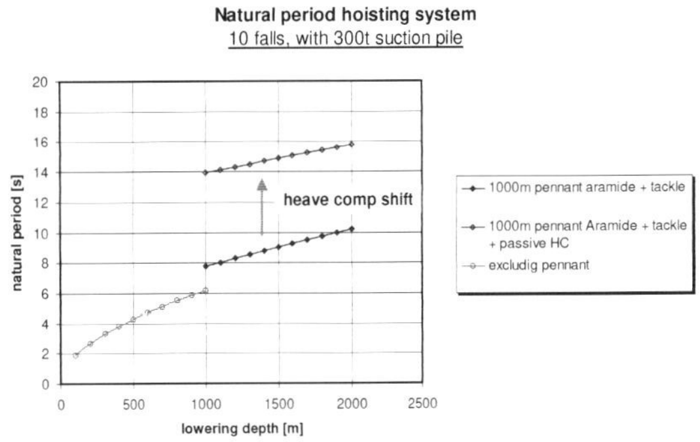

Fig. 12 Heave comp shift

From the figure it can be observed that the natural period of the system including passive heave compensator shifts from 10-11 s towards 14s out of the period of the crane tip motions at a lowering depth of 2000m.

This passive heave compensator is added to the Orcaflex model as well and simulated fore case 1 in the same equivalent wave condition Hs=2 m, Tz=11 s.

The dynamic amplification for this combination decreases from 52% without the passive heave compensator towards below 5% with the passive heave compensator.

CONCLUSIONS

- With the Polyester or alternative fiber mooring line as lift pennants (which have a neutral buoyancy), heavy suction anchors and other offshore structures can be installed in up to more or less unlimited depths.
- The self-weight of the steel hoisting tackle of the crane is now limited to the Cap. depth range of the cranes (approx. 1000 tons at 1000 m depth). Beyond that depth we do not have the self weight disadvantage of the steel wire rope of the hoisting tackle anymore.
- Total depth range and all installation steps should be assessed on natural period and dynamic amplification.
- Resonance problems can be avoided by selecting the correct stiffness of the heave compensator / pennant combination.
- This method can also be used with steel anchor chain or wire rope as pennant, except maximum installation weight is now reduced significantly due to own weight of steel pennant.

REFERENCES

[1] Krabbendam, R.L: Transport and installation of 14 mooring piles at 1300m water depth with a DP 2 heavy lift vessel, December 2009.

[2] Faltinsen, O.M.: Sea loads on ships and offshore structures, Cambridge ocean technology series, Cambridge university press 1990.

[3] DNV-RP-H103: Modelling and analysis of marine operations, draft, December 2008.

[4] R.R.T. van Dijk: Javelin hook motion measurements, Marin report No.21981-1-TM, August 16, 2007.

SESSION D PAPERS

(This page intentionally left blank.)

Proceedings of the 16th Offshore Symposium, February 9, 2010, Houston, Texas
Texas Section of the Society of Naval Architects and Marine Engineers

MODEL TESTS AND SIMULATIONS ON CIRCULAR SHAPED FPSO WITH DRY TREE SOLUTIONS

Marcos Cueva, Fernando Faria
Oceanica Offshore

Arjan Voogt
MARIN

Nico Vandenworm
SSP Offshore

ABSTRACT

New build Floating Production Storage and Offloading units (FPSOs) are traditionally shaped like a shoe box with flared bow to simplify the construction, while taking advantage of existing knowledge from sailing vessels and tankers. In harsher environments these floaters require a turret to rotate into the direction of the main environment. The need for an expensive turret can be eliminated with a circular shaped floater, which also requires less steel per storage capacity. SSP Offshore Inc developed two of these floaters, the SSP and the SSP PLUS. Both units were model tested at LabOceano in Brazil to confirm the calculated motion response and mooring loads.

This paper discusses the specific advantages of both platforms and presents the numerical and experimental models prepared for motion evaluation. The test results demonstrate the importance of second order effects for the pitch motions and confirm the ability to use the SSP PLUS with dry tree risers in Brazilian conditions.

INTRODUCTION

SSP Offshore Inc. developed a family of round floaters which combine small motions with large storage capacity and low steel weight. In this process different companies were contracted to perform specialized evaluations. The current paper discusses the studies related to the floaters motions. Both numerical and experimental models were built and will be presented.

The SSP and SSP PLUS can be classified as mono-columns. These platforms have been studied for several years, Miyagawa *et al* (1989) presented in 1989 a comparison of 4 new floating units, and concluded that the round shape was one with the biggest advantages.

Others, like Matsuura *et al* (1995), Tanabe and Matsuura (1997) and Cueva *et al* (2005), and Cueva *et al* (2006), also made contributions in this floater concept, answering several doubts usually raised by the offshore engineers.

In this paper, an overview of the unit will be presented, followed by details of the numerical model and model tests. The compared results and the conclusions are presented at the end.

SSP AND SSP PLUS OVERVIEW

Both SSP floaters are circular in shape, which makes the unit indifferent for the incidence angles of the environments. Neglecting the asymmetries in the mooring or wind loads, no excitation for yaw exist and the surge and pitch motions equal the sway and roll motions. Therefore the motion response is classified with three types of responses: vertical (heave), horizontal (surge/sway) and angular (roll/pitch).

The SSP consist of a bowl shaped hull with centre column. The roll/pitch natural period of the SSP can be moved outside the range of expected wave frequencies with the design of the weight in the bottom of the centre column. The centre column base is enclosed to bind a significant amount of water which results in added inertia for the roll/pitch motions. The added mass at the bottom of the centre column, also gives a pitch moment due to the surge loads on this segment. This effect shifts the centre of rotation further down below the waterline and couples the surge and pitch response. This results in an increase of pitch motions and available freeboard. The SSP requires approximately 29,000 tons of steel to allow storage of about 1.25 Mbbls, which is significantly more efficient than conventional new built/converted FPSOs.

To improve the heave motions of the SSP and eliminate the need for a centre column, the SSP PLUS compromises some of the steel weight storage ratio with a permanently ballasted skirt below the SSP. Combined with large bilge keels at the bottom the natural period for heave increases to 23 seconds which is well above most of the wave periods. The unit has storage capacity of 1.6 Mbbls of oil, and can support topside up to 20.000t. In wave conditions similar or smaller than Brazil's Santos basin, the unit can facilitate dry tree solutions with proven heave compensator technology.

As a general design for GoM, with 1.000m of water depth, the unit uses a 3x6 mooring arrangement, in a semi-taut layout. The 150mm diameter chains are placed in the top and bottom of the lines, with length of 50m and 100m, respectively. The 312mm polyester line is the mid section, with 1.250m length.

Five SCR risers (varying from 16"~18") were connected at the same elevation as the mooring systems and ran in a region between 2 mooring line groups. In an offshore application, the SCR risers and/or production risers can also be connected at the center/moonpool area of the SSP Plus keel, but the more severe option was tested to comply with earlier studies and check the feasibility of this approach for WoA conditions. The previous study contained 8 SCRs, but the value can vary according to the field characteristics.

Figure 1 – Overview of the SSP Plus Concept

NUMERICAL CALCULATION

To predict the motions of the unit, a numerical model was prepared according to the same data used in the tests.

The diffraction/radiation analysis was performed with the software DIFFRAC, and the time domain coupled analysis, with mooring lines and risers included, evaluated in AnySim, both developed by MARIN.

In order to consider the viscous damping in the analysis, not calculated by potential methods, an external damping of 2% of critical damping was included on all modes of motions.

The mesh used in the diffraction method and the bird view of the complete system, is presented in Figure 2.

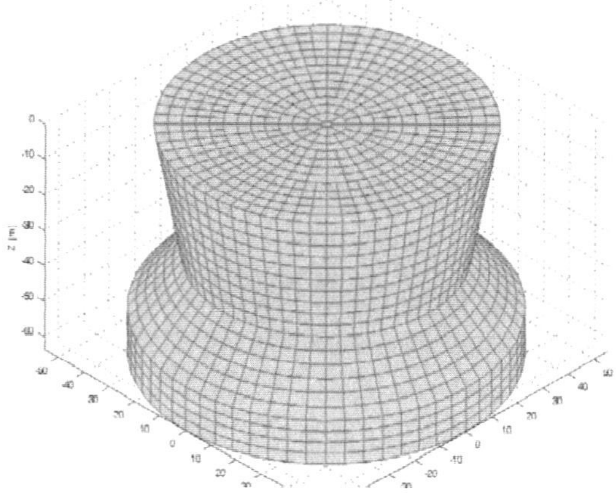

Figure 2 – Mesh for diffraction calculation

MODEL TESTS

In order to verify the motions obtained from numerical analysis, model tests were executed in the LabOceano basin, in Brazil, with specification from MARIN – Maritime Research Institute Netherlands and supervision of Oceanica Offshore, a Brazilian engineering and consulting company.

The facility is composed by a 40m long, 30m wide and 15m deep tank, with waves and wind generation capabilities. The current generation is under final phase of installation and was not generated in the tests.

The model was constructed in the scale of 1:60, with dimensions, inertia, displacement and center of gravity according to the data presented in Table 1. The model is presented in Figure 4

The differences between the specified and obtained data were maintained according to the required by ITTC, less than 1% in the displacement, less than 2% in the vertical center of gravity position and less than 3% in the radius of gyration.

Figure 3 – LabOceano basin

Table 1 – SSP Plus Data

SSP+ CHARACTERISTICS		
Description	**Symbol**	**Value**
Diameter in waterline	Lwl (m)	96.60
Maximum diameter	Loa (m)	121.00
Moonpool diameter		17.00
Depth	H (m)	78.10
Draft	T (m)	59.40
Displacement	Δ (t)	450871.00
Logitudinal pos of CG*	XCG (m)	0.00
Transverse pos of CG*	YCG (m)	0.00
Horizontal pos of CG*	ZCG (m)	26.00
Logitudinal radius of gyration	Rxx (m)	42.00
Transversel radius of gyration	Ryy (m)	42.00
Horizontall radius of gyration	Rzz (m)	45.00
Metacentric height	Gm (m)	8.80

*from center lines and keel

The mooring system tested is a 3x3 semi-taut in 875m WD, with equivalent restoring of a 3x6 system in 1.000m WD. Flexible risers were tested are in free catenary configuration. The bird view of the truncated lines arrangement is presented in Figure 5.

The horizontal restoring stiffness was adjusted with maximum difference of 6%.

The objective of the test was to verify different conditions in the main offshore areas in the world. The waves presented in Table 2 were calibrated. The environmental conditions cover a large range of wave periods and wave heights. The different wave heights show the linearity of the motion response, while the large range of periods ensures that all motion phenomena are covered in the tests.

Figure 4 – Overview of the SSP Plus model

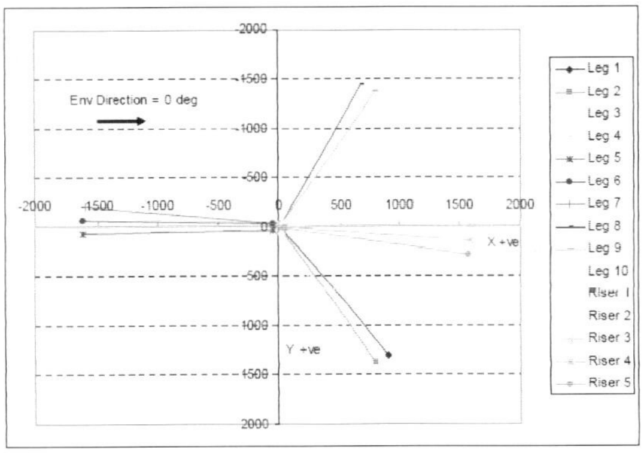

Figure 5 – Truncated mooring and riser system

It should also be noted that the most severe conditions for the mentioned locations are included. The 100 yr condition in the Campos basin in Brazil for instance is only 7.8m, compared to the tested 11.11m in the Santos basin. Further the latest Gulf of Mexico hurricane conditions are used for the central region and even a 1000yr wave is considered with significant wave height of 19.8m.

The system was tested in the conditions presented in Table 2 and the results analyzed both in time domain and in frequency domain.

Table 2 – Environmental conditions calibrated

	WAVE		WIND
CONDITION	Hs(m)	Tp (s)	Vw (m/s)
Brazil 1yr	5.70	13.70	19.00
Brazil 100ya	9.87	11.00	30.00
Brazil 100yb	10.79	13.50	30.00
Brazil 100yc	11.11	16.00	30.00
Brazil 100yd	10.35	18.50	30.00
WoA 1y	2.70	15.00	19.00
WoA1oy	3.40	17.00	25.00
WoA 100y	3.80	18.00	31.00
GoM 1y	4.90	11.10	14.50
GoM 100y	15.8	15.40	48.00
GoM 1000y	19.8	17.20	60.00

In addition to motion response tests in operational and survival conditions, offloading simulations were carried out in the operational conditions and in squalls. The squalls were simulated with increasing wind perpendicular to the waves, so that the tanker was forced to rotate around the SSP. To facilitate this rotation a special reel system was included, shown in Figure 6. Some fishtailing occurred in the collinear conditions, which was suppressed with stern thrust on the tanker, simulating a tug pulling controlling the vessel at the stern. The response to the squall condition showed promising results and will be further studied with tug master and marine specialist with several operations in a Joint Industry Project (JIP). The results are not further discussed in this paper, but will be shared in this JIP.

Figure 6 – Reel system which allows the shuttle tanker to weathervane around the SSP+

RESULTS

The results for both numerical and experimental evaluations are presented in terms of RAOs, for the heave and pitch/roll motions, and in terms of standard deviation and most probable maxima in three hour simulations.

First Figure 7 presents the RAOs for pitch derived from all the tests, showing a good agreement between numerical and experimental for the first order response. In the large range of environmental conditions tested, the response is relatively linear (all colored lines coincide).

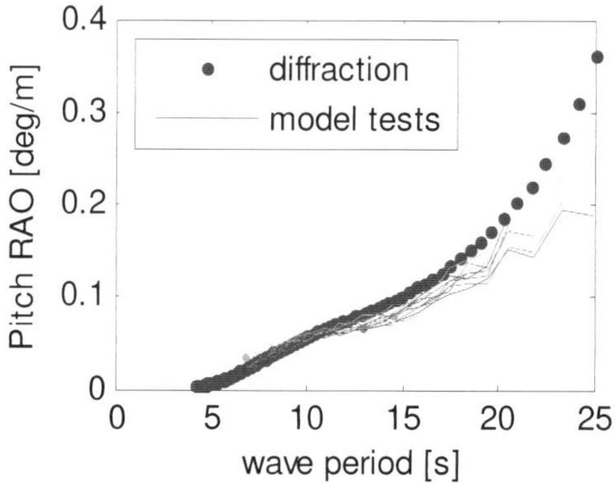

Figure 7 – Pitch RAO: Numerical x Experimental

The pitch response in the model test is slightly smaller than predicted by the simulations. The decay tests (Figure 8) shows that the damping is much larger than in the initial simulations and is mainly quadratic. The dotted line is the same in both graphs and represents the measurement. The solid line is the numerical result in the upper plot and the best quadratic fit on the model tests in the lower plot.

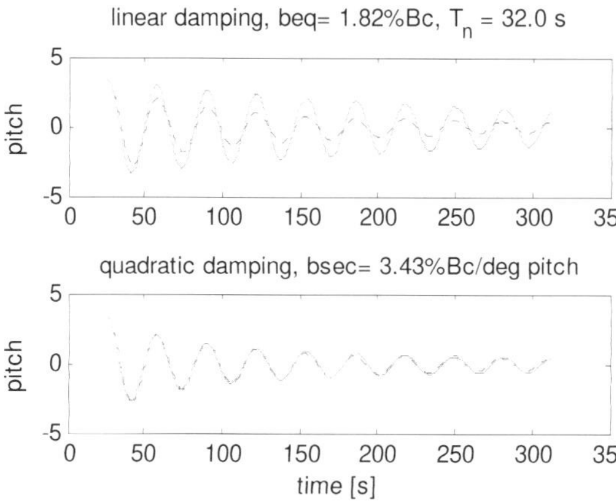

Figure 8 – Pitch Decay

The natural periods are high: 23s for heave and 32s for pitch, resulting in small response in to the first order waves, helping the riser design both in extreme and fatigue criteria.

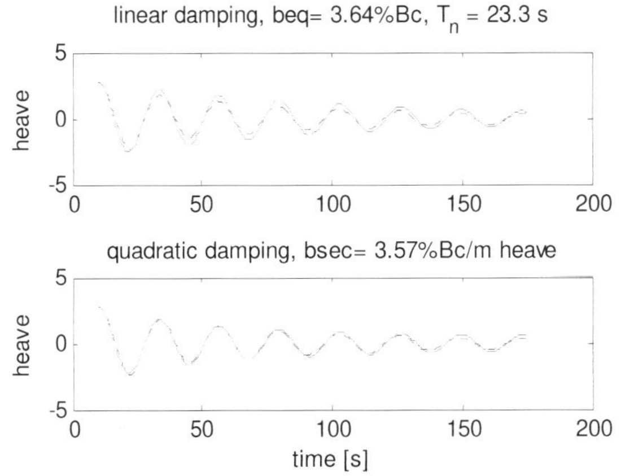

Figure 9 – Heave Decay

The heave RAO is also fairly linear for wave periods shorter than the natural period. Figure 10 shows the RAO derived from the waves with peak period above 16 seconds. The response at the natural frequency is dependent on the wave height. This is induced by the viscous damping (see decay test in Figure 9), resulting in a smaller heave response for larger waves due to vortices shedding of the bilge keels. As a result the diffraction analysis shows good comparison for the operational sea states, but overestimates the motions in storm conditions.

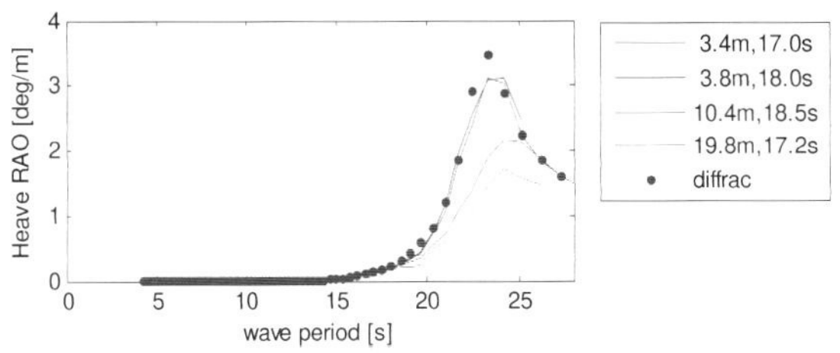

Figure 10 – Heave RAO

Figure 11 compares the heave response at the centre of gravity in beam seas with other production floaters, showing a significant improvement of the response compared to traditional ship shaped FPSOs.

In Table 3 the standard deviations are presented for heave and pitch. The comparison with numerical simulations shows that the numerical simulations are conservative for the heave motions in storm conditions, in line with the observation for the RAO. However the pitch motions are underestimated in the calculations, although the damping in the tests was larger than assumed in the calculations.

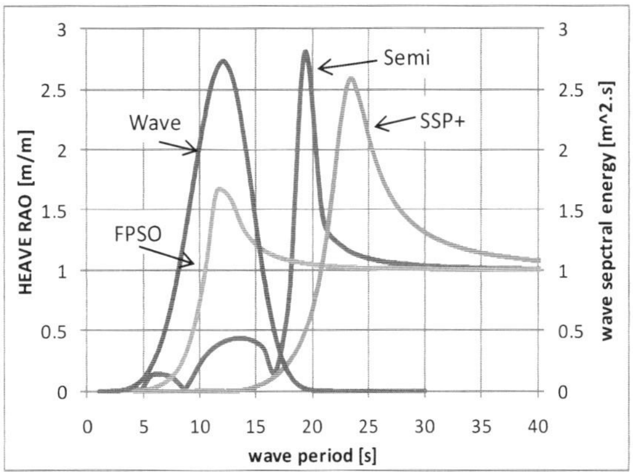

Figure 11 – Heave RAO compared to other floaters

Table 3 – Motions statistical values

St. dev.	numerical results		model tests values	
Condition	Heave [m]	Pitch [deg]	Heave [m]	Pitch [deg]
Brazil 1yr	***0.26***	0.13	0.16	***0.30***
Brazil 100ya	0.06	0.31	***0.19***	***0.73***
Brazil 100yb	***0.43***	0.28	0.33	***0.90***
Brazil 100yc	***1.89***	0.36	0.71	***1.04***
Brazil 100yd	***3.41***	0.48	1.45	***0.90***
WoA 1yr	***0.23***	0.07	0.12	***0.09***
WoA 10yr	***0.64***	0.12	0.36	***0.14***
WoA 100yr	***0.95***	0.15	0.51	***0.17***
GoM 1yr	0.02	0.09	0.09	***0.25***
GoM 100yr	***2.03***	0.51	0.73	***1.71***
GoM 1000y	***4.66***	0.80	1.46	***2.68***

The spectral content of the pitch motions (see Figure 12) shows that significant motions occur around the natural frequency. Comparison with the numerical model which only includes first order motions, shows an under prediction of the second order response.

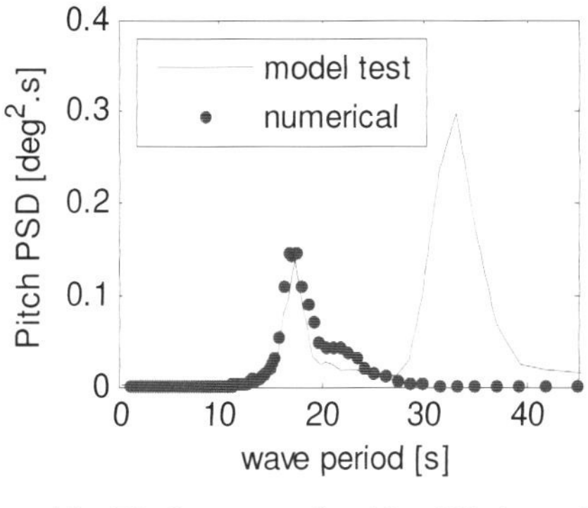

Figure 12 – Pitch spectra for 10yr WoA condition

This second order roll and pitch phenomena is observed before on drilling semi submersibles with low metacentric height (see Voogt (2002). Due to the underwater hull shape second order excitation exists and with the very small pitch response at the wave frequencies, these forces dominate the response spectrum. The study on semi submersibles shows, motions are smaller for shorter natural periods. The pitch natural period can be reduced by relocating the permanent ballast further towards the centre. This can further reduce the pitch motions in the GoM conditions, but for the generic unit this was not an option since it would increase the first order response in long period swells West of Africa.

A weibull analysis is used to analyze the extremes and determine the most probable maxima in a 3 hour storm. The resulting values are shown in Table 4.

Table 4 – Most probable maximum motions

Tests	Heave [m]		Pitch [deg]	
Condition	MPM-	MPM+	MPM-	MPM+
Brazil 1yr	-0.56	0.53	-0.95	1.32
Brazil 100ya	-0.57	0.62	-2.82	2.10
Brazil 100yb	-1.13	1.09	-3.22	3.39
Brazil 100yc	-2.36	2.38	-3.92	4.25
Brazil 100yd	-4.68	4.71	-3.18	4.07
WoA 1yr	-0.36	0.35	-0.38	0.26
WoA 10yr	-1.22	1.24	-0.59	0.39
WoA 100yr	-1.65	1.65	-0.67	0.60
GoM 1yr	-0.29	0.31	-0.91	0.76
GoM 100yr	-2.36	2.35	-6.89	7.48
GoM 1000y	-4.88	5.19	-10.14	11.05

The largest heave motions are observed in the 1000 yr Hurricane condition and are comparable to the heave response in the 100yr Brazil condition with the 18s wave period. It should be noted that even though these conditions are extremely severe, dry tree production is possible with 20m stroke heave compensators, which are available.

The pitch motions are largest in the hurricane conditions. As discussed these motions are dominated by second order effects and are comparable to the response of drilling semi submersibles in these conditions. To further study the measured pitch response, Figure 13 compares the most probable maximum pitch with the significant wave height in each condition. The tests with wave peak periods between 15s and 17s are marked with green triangles, while the remaining test results are marked as diamonds. The effect of the wave period seems limited in the tested range, indicating that the pitch response is not driven by resonance at twice the wave period, but by the drift forces resulting in a quadratic increasing pitch response with increasing wave height (the red line indicating a fit equal to MPM=0.28.Hs^2) .

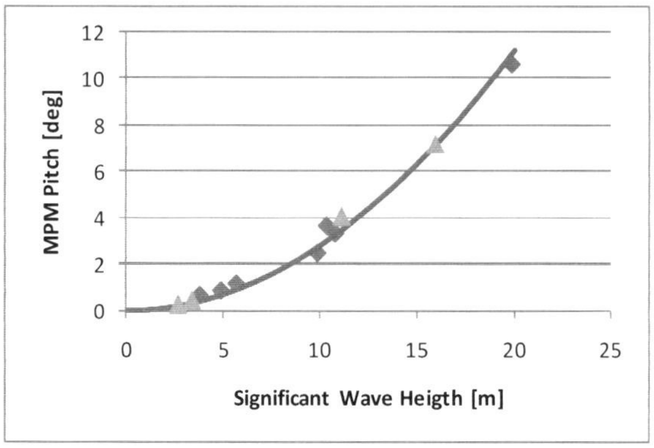

Figure 13 – MPM Pitch as function of the significant wave height

When the risers are connected at the outside of the unit, the fatigue live will be influenced by the pitch motions, but this can be prevented by running the SCR's through the centre column. The pitch motions also reduce the available freeboard, resulting in possible green water on deck which needs to be deflected or prevented by raising the freeboard.

Overall the motions are very limited for almost all conditions, and the environmental conditions with larger motions are extreme, and should be analyzed considering this fact.

Table 5 – Mooring lines tension and offsets

	Numerical Values		Experimental values	
CONDITION	Surge St.Dev [m]	Max ML Load [kN]	Surge St.Dev [m]	Max ML Load [kN]
Brazil 1yr	1.58	2549	2.00	3214
Brazil 100ya	10.85	7928	8.57	6340
Brazil 100yb	6.02	5604	6.83	5523
Brazil 100yc	4.97	5162	5.97	5093
Brazil 100yd	5.78	5765	4.64	5060
WoA 1y	0.46	2073	0.77	2734
WoA10y	0.61	2337	0.86	2889
WoA 100y	0.82	2617	1.41	3083
GoM 1y	1.87	2486	2.30	3218
GoM 100y	9.86	9403	9.87	8191
GoM 1000y	14.2	14576	10.00	12239

The extreme mooring tensions are presented in Table 5, with the standard deviation of the surge motions. Each model test line is equivalent to two mooring lines in real scale and the maximum top tensions observed are below the required limit by API. Further, the standard deviation of the surge motions is less than 1.5% of the tested water depth.

CONCLUSIONS

The SSP+ was tested in a very broad range of waves, which can possible represent almost all conditions that can occur offshore. Special attention was dedicated to the harsh conditions.

The model tests confirmed the large natural periods and showed motion characteristics similar to a SPAR. The heave motions are significantly reduced compared to conventional FPSOs, allowing dry tree applications in a broad range of environments.

The vertical motions are limited to motions around the natural frequencies. Decay tests show that a significant amount of viscous damping exists which is quadratic in nature. The influence of the tested mooring system on the heave and pitch decays is limited.

The heave motion RAO show a reduction of the amplification at the natural frequency for increasing motions. This phenomenon is induced by the second order damping and results in a less than linear increase of the heave motions with the wave height.

Due to second order excitation the pitch extremes increase with the wave height squared. In all non-hurricane conditions the extreme pitch motions are less than 4 degrees.

REFERENCES

[1] Cueva, M., Malta, E.B., Nishimoto, K., Costa, A.P., (2005), *Estimation of Damping Coefficients of Moonpools for Monocolumn Type Units*, OMAE 2005-67332;

[2] Cueva, M., Fujarra, A.L.C., Nishimoto, K., Quadrante, L., Costa, A.P., (2006), *Vortex-Induced Motion: Model Testing of a Monocolumn Floater*, OMAE 2006-92167;

[3] Matsuura, M., Isozaki, Y., Ishibashi, Y., (1995), *Development of Mono-Column Type Hull Form with Passive Type Motion Damping Devices for Floating Production System*, OMAE 1995, Volume I-A, pgs. 403-410;

[4] Miyagawa, M., Isozaki, Y., Yano, S., Matsuura, M., (1989), *Development of New Concepts on Floating Production storage and Offload Unit*, OMAE 1989, Volume I, pgs 49-55;

[5] Tanabe, A., Matsuura, M., (1997), *FPS Development – A New Concept of MONOCOLUMN*, Black SEA-97;

[6] Voogt A.J., J.J.Soles and R.v.Dijk.,(2002), *Mean and low frequency roll for semi-submersibles in waves*, ISOPE 2002, Kyushu, Japan, 2002.

(This page intentionally left blank.)

Proceedings of the 16th Offshore Symposium, February 9, 2010, Houston, Texas
Texas Section of the Society of Naval Architects and Marine Engineers

EFFECT OF BILGE KEEL GEOMETRIES ON ROLL RESPONSE OF FPSO HULL FORMS IN WAVES

Y.H. Yu and S.A. Kinnas (M)
Ocean Engineering Group, Department of Civil, Architectural and Environmental Engineering, The University of Texas at Austin, Austin, Texas, USA

ABSTRACT

A study on the hydrodynamic roll response of typical FPSO and LNG hull-forms near resonance in beam seas is presented. A 2-D finite volume method-based Navier-Sokes solver is applied to analyze the flow around the hull sections. The effect of hull geometries and the bilge keel widths on roll-decay motion as well as on transient roll response of hulls in regular waves for a range of wave frequencies is analyzed.

KEY WORDS

Navier-Stokes (NS) solver; Finite Volume Method (FVM); FPSO; LNG; ship-shaped hull; bilge keels; free surface

INTRODUCTION

Floating, Production, Storage and Offloading (FPSO) drilling vessels, often converted from tankers, are a typical type of floating production system, used for the extraction of oil in deep seas. FPSO based offshore systems have gained importance in recent years. However, FPSO hulls have often been found to be subject to excessive roll motions, due to the prevailing harsh weather conditions in deep seas, which inhibit their use as stable production platforms. Bilge keels are passive stability devices and are implemented to reduce the roll of hulls. Bilge keels have the disadvantage of increasing the hydrodynamic resistance of the vessel. Nevertheless, it is an effective and economic way of mitigating roll motions of stationary. Their effectiveness lies in their ability to damp out roll motions over a range of wave frequencies.

Vugts (1968) made a great contribution in investigating independent roll, heave and sway motions for various hull geometries. He was the first to measure and calculate the hydrodynamic added-mass and damping coefficients, which were conventionally used for evaluating the performance of a hull. His study also pointed out the importance of the effect of viscosity. Over the years, numerous numerical simulations and experimental measurements were focused on evaluating these coefficients for various hull models, bilge keel geometries and bilge keel widths (Na et al. 2002; Roddier 2000; Wilson et al. 2006; Yeung et al. 1998; Yeung et al. 2000; Yuck et al. 2003). These studies showed that strong flow separation was observed near the sharp edge of the barge, and the wave propagation on free surface carries energy, contained by the shed vortices, away from the hull. Downie et al. (1988) studied the roll response of ship hulls near resonance and modeled vortex shedding from the bilges of a barge hull of rectangular cross-section via using a discrete vortex method. Their study showed that the damping for a sharp–edge barge is greater than the one with round edges, which leads to the conclusion that vortex shedding is particularly important to roll damping. Several studies also focused on the prediction of transient responses of hulls in waves (Esperança et al. 2008; Jung et al. 2005; Jung 2004; Roddier 2000; Wilson et al. 2006). In particular, Jung (2004); Jung et al. (2005) measured the flow characteristic around a rectangular barge in a 2-D wave tank via using a particle image velocimetry system. Their study found that those vortices around the barge seemed to excite, instead of reducing, the roll motion when the wave period is larger than the natural roll-decay period. Esperança et al. (2008), performed an experimental study of roll motions on an FPSO, where the effectiveness of using bilge keels on roll damping were demonstrated, and the bilge keel loads were evaluated from the bilge keel strains via using plate theory.

A 2-D unsteady incompressible NS solver (NS-2D) was developed by the Ocean Engineering Group at the University of Texas at Austin (Yu 2008). A series of studies on the hydrodynamic performance of hulls, with an emphasis on roll, were performed via using NS-2D (Kinnas et al. 2007; Yu 2008; Yu and Kinnas 2008; 2009b; Yu et al. 2005). The numerical solver was tested and validated in prescribed roll and also in transient-decay motions through the comparison of NS-2D simulations and other numerical and existing experimental results (Vugts 1968; Yeung et al. 2000; Yuck et al. 2003). The NS-2D solver was then applied to predict the roll response of hulls in regular wave (Yu and Kinnas 2009a) for a range of wave frequencies , and the bilge keel loads on the up-wave and down-wave sides of the hull were successfully examined.

The objective of this work is to study the roll response of ship-shaped hulls near resonance in beam seas using a 2-D Navier-Stokes solver. The properties and the geometries of the hull and the bilge keels widths are given based on the experimental study of Esperança et al. (2008). NS-2D is applied to model the flow around the hull sections and the interaction between waves and the barge. A study on the roll-decay motion as well as on the transient roll response in regular waves for a range of wave frequencies is presented. The effect of different hull corner geometries and the effectiveness of using bilge keels on roll reduction are studied.

METHODOLOGY

Governing Equations and Discretization

A cell centered-based FVM is applied in NS-2D. The discrete representation of the continuity equation and the conservative form of the Navier-Stokes equations are given as:

$$\sum_{\partial\forall} \rho(\vec{U}_f \cdot \vec{n}) A_f = 0,$$
$$\frac{\partial \vec{U}_f}{\partial t}\forall + \sum_{\partial\forall}\left[v_n \vec{U}_f - \nu\left(\frac{\partial \vec{U}_f}{\partial n}\right)_f\right] A_f = -\frac{1}{\rho}\sum_{\partial\forall} p_f \vec{n} A_f, \tag{1}$$

where $\forall$ represents the cell area, $\partial\forall$ denotes the boundary of the cell, v_n is the normal velocity on the cell face, and A_f is the width of the cell side. $\partial/\partial n$ indicates the derivative with respect to the normal direction of the cell face, and $\vec{n}$ is the unit normal vector to the cell side. Subscripts "P" and "f" represent the values located at the cell center and at the midpoint of the cell face. $\vec{U}=(u,v)$ represents the flow velocity vector with respect to the inertial coordinate system with (u,v) being the two velocity components, respectively. ρ represents density, p denotes the hydrodynamic pressure, t indicates time, and ν represents the kinematic viscosity.

Roll Response

The FPSO hulls are assumed to be slender with the longitudinal length-scale exceeding the other two physical dimensions. This assumption allows the modeling of the flow around the hull in a 2D strip-wise manner. The hull properties are given based on the experimental settings in Esperança et al. (2008). The hull is presumed to roll about its metacenter M_T, which is $0.0315B$ above its origin O, the intersection of the mean free surface and the longitudinal axis of the hull, when the hull is at rest, where B indicates the hull beam. The center of gravity is located $0.142B$ below M_T, and the radius of gyration in roll is equal to $0.322B$. In this paper, the horizontal and vertical motions are constrained in the study. The orientations of the boundaries and the computational domain $\Omega(t)$ are shown in Fig. 1.

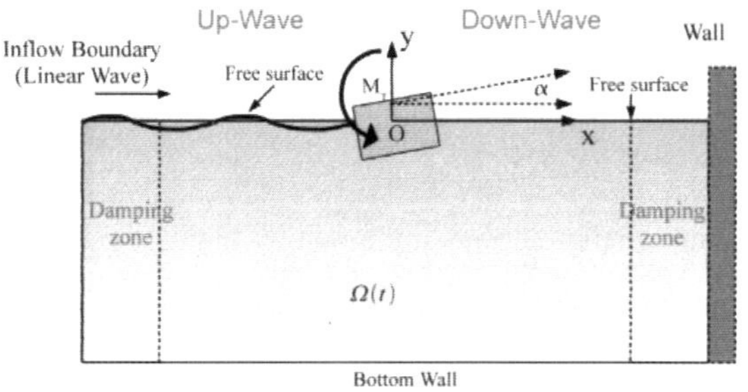

Fig. 1 Fluid domain and the domain boundaries

The response of the angular acceleration $\ddot{\alpha}$ is calculated based on the Newton's second law, and the dynamic equation of roll is given as

$$M_o - W_b x_g = I_O \ddot{\alpha}, \tag{2}$$

where I_o is the mass-moment of inertia about the center of rotation, W_b is the weight of the body, and x_g represents horizontal distance from the center of gravity to the center of rotation. The total moment M_o on the hull are evaluated by integrating the cross product of the elemental force $\vec{F}$ and the position vector $\vec{r}$ on the wetted portion of the body surface,

$$M_o = -\int_{H(t)} \vec{F} \times \vec{r} d\Gamma, \tag{3}$$

where r represents the distance vector from the center of rotation and Γ denotes the hull-surface length. The elemental force contains the forces due to buoyancy *Fb*, pressure *Fp* and friction *Fr*, which are given in tensor notation,

$$\begin{aligned} Fp_i &= pn_i, \\ Fb_i &= -\rho gyn_i, \\ Fr_i &= \rho(U_{i,j} + U_{j,i})n_i. \end{aligned} \quad (4)$$

The roll angle is then calculated by integrating the angular acceleration over time.

Hull and Bilge Keel Geometries

The bilge keel widths and the hull geometries are defined based on Esperança et al. (2008) (a scale of 1:70 model with B=57.2m in full scale). The draft to beam ratio is approximately equal to 0.38. The keel widths for the three bilge keel models (bk1, bk2 and bk3) are equal to 0.0175B, 0.035B and 0.0525B, respectively. Note that the exact hull geometry used in Esperança et al. (2008) is not available. Two types of hull corner geometries are considered in the study. As shown in Fig. 2, one is a round corner with a bilge radius equal to 0.04B, and the other is flat corner with a cut slop equal to 45°.

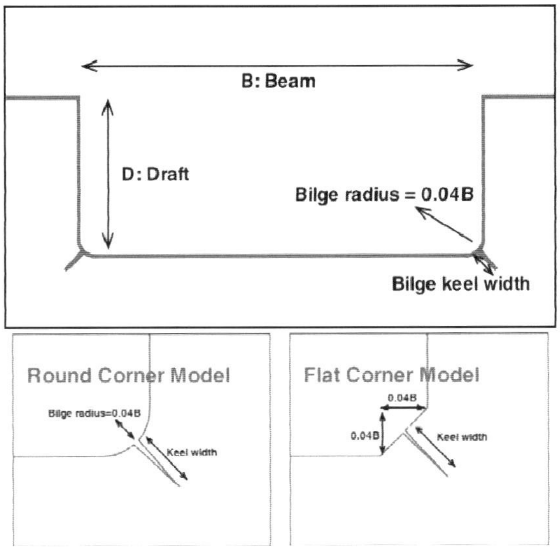

Fig. 2 Shape of hull models

Numerical Algorithms and Settings

The numerical schemes applied in NS-2D are briefly introduced in this section, and more details of the methods are described in Yu (2008).

The momentum equations and the continuity equation are solved based on the primitive variable method. Therefore, the SIMPLEC pressure-correction

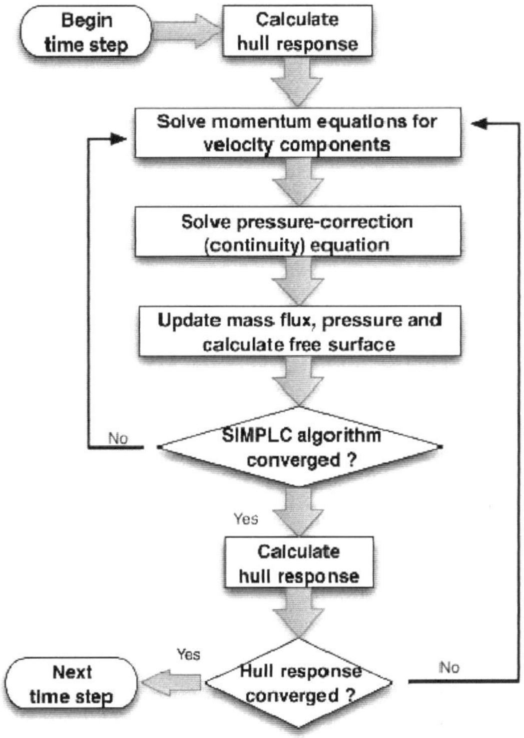

Fig. 3 Overview of the numerical algorithms

algorithm is applied for pressure-velocity coupling, where all the unknown variables, pressure and velocity components, are updated by the correction term through iterations. A 3rd-order MUSCL (Monotone Upstream-Centered Schemes for Conservation Laws) scheme is utilized for handling the convective terms in the momentum equations, and a 2nd-order implicit scheme is used for time marching. A free-surface tracking algorithm is implemented to model the wave propagation. As shown in Fig. 3, the calculation of the momentum and pressure-correction equations, the free-surface elevation and the roll response of hulls are coupled through iterations. The linear sparse systems of the equations resulting from the pressure-correction equation and the momentum equations are then solved using the Generalized Minimal Residual (GMRES) algorithm. Note that the effect of turbulence on predicting the corresponding hydrodynamic loads has been found to be small as reported in Wilson et al. (2006); Yu (2008); Yu and Kinnas (2008). The turbulent flow modeling is therefore excluded in the study.

Moving and deforming meshes are used to follow the hull motion and to track the nonlinear free surface, where re-meshing is applied at every iteration inside the SIMPLEC algorithm. The cell-face motion terms are therefore added to the momentum equations, and the

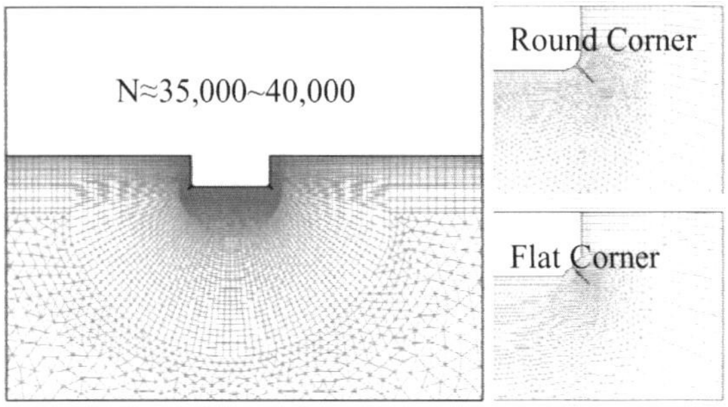

Fig. 4 Details of the grid near the hull

momentum equations must satisfy the Geometric Conservation Law (GCL).

The total number of elements for each hull model is in the range between 35,000 and 40,000, and the details of the grid near the hull and its corner are presented in Fig. 4. Prism layer cells are placed along the hull surface, and the mesh is mixed with triangle and quadralateral cells. The minimum cell height for the prism layer cell is approximately equal to $0.005B$. The details of the grid-dependence studies were addressed in Yu (2008); Yu and Kinnas (2008). All the studies in this paper are performed using 2.4 GHZ quad-core 64-bit Intel Xeon processors, with 2 GB Memory per core. It takes about 12 hrs on a single core to complete 6000 time steps in roll-decay test and 12 hrs to complete 2500 time steps for simulating the roll response of a hull in waves.

Free Surface Tracking

The non-linear boundary conditions are applied on the free surface, where the Kinematic Free Surface Boundary Condition (KFSBC) and the Dynamic Free Surface Boundary Condition (DFSBC) are given as:

$$\text{KFSBC: } v_{FSn} - v_n = 0, \qquad \text{DFSBC: } p_{ap} = p - \rho g \eta, \tag{5}$$

where η represents the wave elevation, g indicates gravity and v_{FSn} is the normal component of the free surface velocity. As shown in Fig. 5, a free-surface

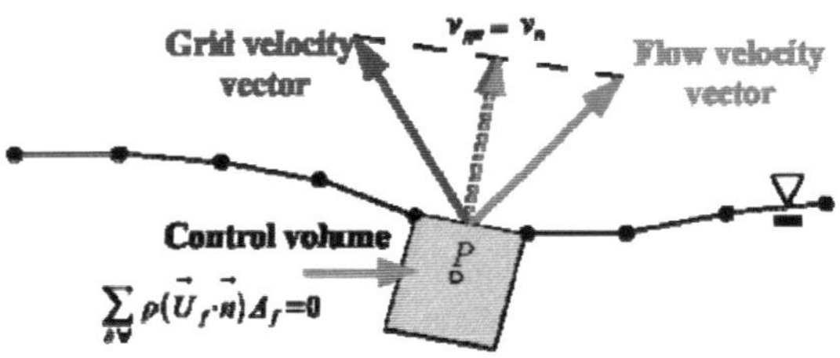

Fig. 5 Boundary conditions for the cell on the free surface and the application of the free-surface tracking algorithm

tracking algorithm is applied based on KFSBC, where the normal grid velocity v_{gn} on the free surface is presumed to be equal to the normal free-surface velocity as well as the normal flow velocity v_n on the free surface, which is calculated by forcing the continuity equation to be equal to zero. The tangential velocity on the free surface is obtained through linear extrapolation, and the pressure value at the center of those cells along the free surface is calculated via Taylor series expansion.

Wave Absorbing

As shown in Kinnas et al. (2007), the sponge layers have been found sufficient to absorb the reflected and outgoing waves. Hence, two artificial wave-damping zones (sponge layers) are placed in front of the up-wave and the down-wave boundaries to reduce the computational domain size. An additional numerical damping term is therefore added to the RHS of the momentum equation.

$$v_d(\vec{U}_B - \vec{U}), \quad \text{where } \begin{cases} \vec{U}_B = \text{linear wave, on the up-wave side,} \\ \vec{U}_B = 0, \text{ on the down-wave side,} \end{cases} \tag{6}$$

$$v_d = \begin{cases} \dfrac{\beta}{2}\left[1 - \cos\pi\left(\dfrac{x - x_d}{x_u - x_d}\right)\right], & \text{in the damping zone} \\ 0, & \text{out of the damping zone} \end{cases} \tag{7}$$

where β is the damping coefficient; x_d and x_u are the downstream and upstream extents of the damping zone. Also, the bottom boundary is presumed to be deep enough to prevent any wave interaction. Hence, on these boundaries, the velocity is presumed to be zero.

BACKGROUND

The performance of a hull in roll is measured, conventionally, in terms of hydrodynamic coefficients, where the hydrodynamic moment of the prescribed roll motion can be written as a linear combination of the added-mass a_{66} and damping b_{66} terms via using Fourier analysis (Na et al. 2002; Roddier 2000; Wilson et al. 2006; Yeung et al. 1998; Yeung et al. 2000; Yuck et al. 2003). Yu and Kinnas (2009b) also demonstrated a very successful correlation between the results from NS-2D and those from the experimental measurement of Yuck et al. (2003), as shown in Fig. 6. The study also showed that the hull geometry has a significant influence on the roll damping performance.

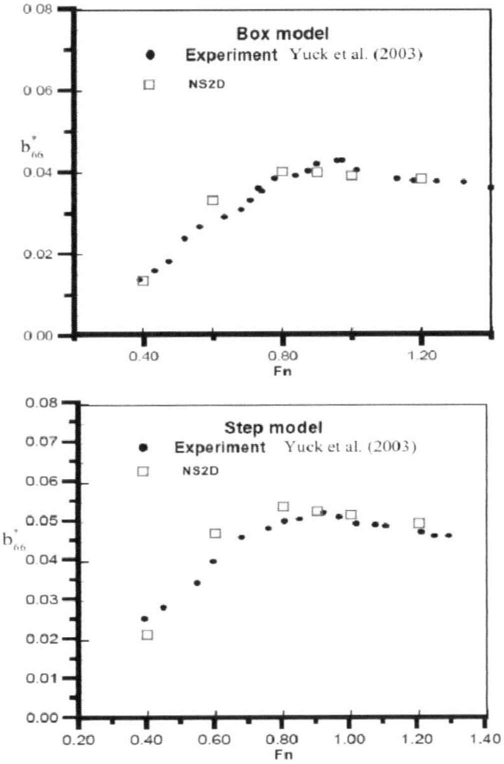

Fig. 6: Comparisons between the damping coefficients (prescribed roll amplitude α_0=5°) from experimental data and those from NS-2D (from Yu and Kinnas 2009b)

RESULTS

In this paper, the study focuses on the wave frequencies that are close to the hull natural roll-decay frequency. Hence, the studies of both the transient roll-decay motions and the roll responses of hulls in regular waves are presented in this section. The results in this paper are all presented in non-dimensional form, where all the variables in NS-2D are made non-dimensional based on B, ρ and g. The non-dimensional time variables and the Reynolds number are defined as

$$t^* = \frac{t}{\sqrt{B/g}},\ T_w^* = \frac{T_w}{\sqrt{B/g}},\ T_n^* = \frac{T_n}{\sqrt{B/g}} \qquad (8)$$

$$\mathrm{Re} = \frac{B^{3/2}\sqrt{g}}{\nu}, \qquad (9)$$

where T_w and T_n indicate the incoming wave period and the natural roll-decay period of the hull, respectively. Re is approximately equal to 2.3×10^6 in NS-2D (based on the model scale in Esperança et al. (2008).

Roll Decay

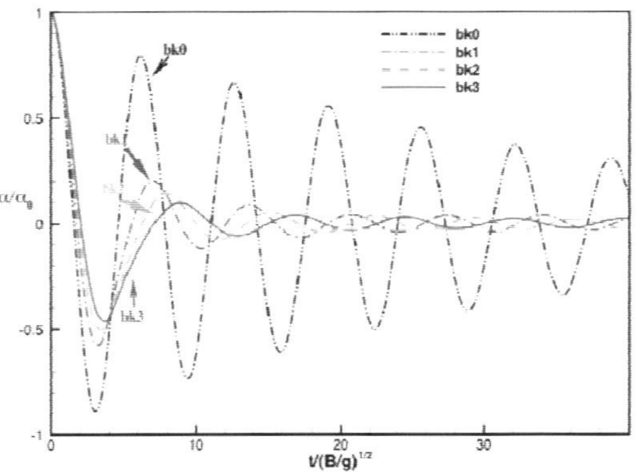

Fig. 7 Roll response for hulls (round corner model) with different bilge keel widths in roll-decay test

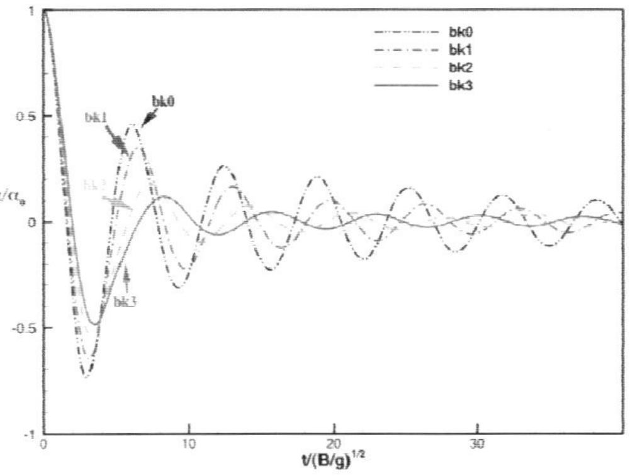

Fig. 8 Roll response for hulls (flat corner model) with different bilge keel widths in roll-decay test

The roll-decay test is performed with a given initial roll angle of α_0=14.32°, and the time step size is equal to Δt^*=0.01. The up-wave inflow boundary is replaced by a wall boundary in the roll-decay test. The simulations are terminated at t^*=60, before the radiated waves reach the outer boundaries. Figs. 8 and 9 present the roll response histories for hulls with two different corner geometries (round and flat) and with different bilge keel widths. The values of the natural roll-decay period are averaged from the first five periods, and they are listed in Table 1. As expected, the roll-response amplitude is reduced when the bilge keels are installed. Both the rate of roll reduction and the roll-decay period are increased with an increase in the bilge keel width, as also mentioned in Esperança et al. (2008). The effect of the hull corner geometry is also important to roll damping. When the bilge keel is not installed, roll damping increases for the flat corner hull as compared to the round corner hull. On the other hand, since the distance between the metacenter and the root of the bilge keel is shorter for the flat corner model, as compared to the round corner model, those sharp-edge induced shed vortices are closer to the hull in the flat corner hull cases. In order to provide same damping performance, a longer keel is therefore required for the flat corner model.

Table 1. Averaged Natural Roll-decay periods

Keel Width	Averaged Roll-Decay Period $T_n/(B/g)^{1/2}$	
	Round Corner	Flat Corner
0.0000B (bk0)	6.43	6.35
0.0175B (bk1)	6.80	6.60
0.0350B (bk2)	7.23	6.85
0.0525B (bk3)	7.91	7.42

Roll Response in Regular Waves

An investigation of roll responses for hulls in waves is performed using the exact same hulls that were analyzed in the roll-decay test. The study is performed in a 2-D strip-wise manner, and a linear wave profile with a wave height of H/B=0.05 is given at the up-wave inflow boundary. The time step size is determined based on the wave period, where approximately 126 time steps are given per wave period. Therefore, the time step size varies from Δt^*=0.04 to Δt^*=0.065 to cover the range of incoming wave periods from T_w^*=5.03 to T_w^*=8.17. The domain size is also adjusted for different incoming wavelengths. The up-wave and down-wave boundaries are approximately 8 and 10 wavelengths away from the hull, respectively. As shown in Fig. 9, the grid dependence study is performed to determine the necessary number of cells per wavelength. As a result, 40 cells per wavelength mesh is used in the following study.

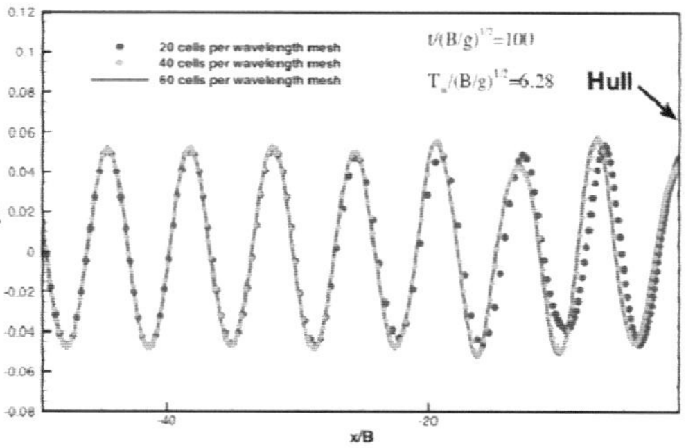

Fig. 9 Grid dependence study of number of cells per wavelength

In order to save computational time, Pre-existing wave and velocity profiles (approximately 7 wavelengths long), as shown in Fig. 10, are given as the initial solution to start the simulation. Fig. 11 shows an example of the wave field on the up-wave and down-wave sides of the hull. An increase on the wave height out of the damping zone and an irregular wave profile near the hull are observed on the up-wave side, which are attributed to wave reflections, and the outgoing wave is successfully absorbed by the damping zone on the down-wave side.

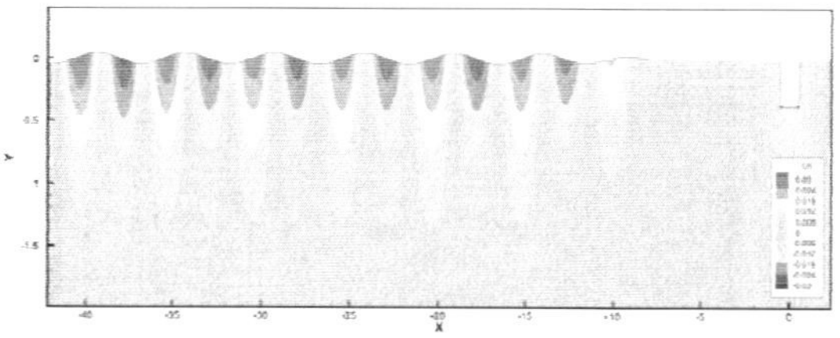

Fig. 10 Initial wave field and x-velocity component contour on the up-wave side of the hull.

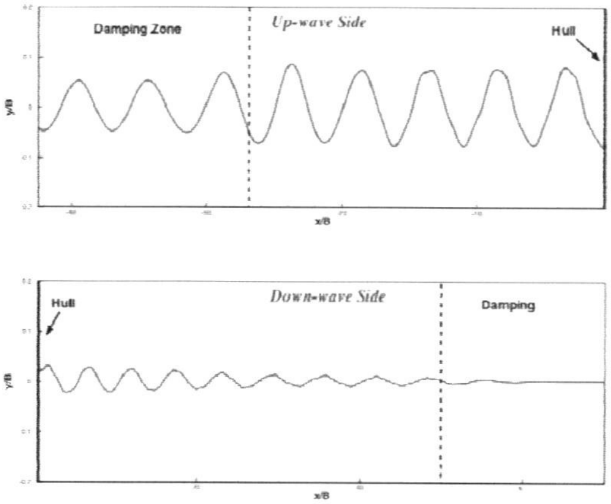

Fig. 11 Wave field on the up-wave and down-wave sides of the hull (bk2 and T_w*=6.28)

Figure 12 shows the roll response histories for hulls with round and flat corners and with different bilge keel widths in regular waves (T_w*=5.65 and 6.91). Note that the roll-response amplitude varies from period to period, particularly for bk0, which indicates a more complex interaction between waves and the floating-hull motions. The roll damping increases when the bilge keel is installed, and the roll reduction is increased as the keel width increases. In the case of T_w*=6.91, the study also shows that a hull with different bilge keel widths can result in a phase shift of the roll response, which implies that the roll response seems to be more dominant by the roll decay motion when the incoming wave period increases.

Round Corner Model Flat Corner Model

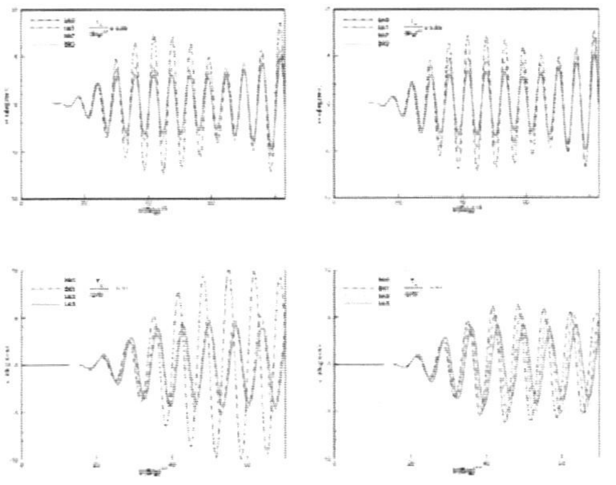

Fig. 12 Roll response for hulls with different bilge keel widths (T_w^*=5.65 and 6.91)

Vorticity magnitude contours are plotted in Figs. 13 and 14. The position for each hull is close to its roll maximum. The shed vortices are observed near the hull when the bilge keels are installed as well as for the flat corner model even when the bilge keels are not installed. The bilge keel induced shed vortices are located further away from the hull for the round corner model, as compared to the flat corner model, which allows more energy to be carried away by the wave and results in a better roll reduction performance.

Figures 15 and 16 present the maximum roll amplitude over a range of wave periods for hulls with round and flat corners and with different bilge keel widths. Note that the maximum and minimum values in the roll response history are used to determine the heights of the variations, and the maximum amplitude is evaluated as half of the maximum value of those heights. As expected, the peak value is observed near resonance ($T_w = T_n$). The peak value drops significantly as the bilge keel is installed, and the reduction is increased when the bilge keel width increases, as reported in Esperança et al. (2008). The result shows that the bilge keels are effective on roll reduction for a wide range of wave frequencies. Within this range of wave frequencies and with same bilge keel width, the roll response for the round corner hull is smaller than

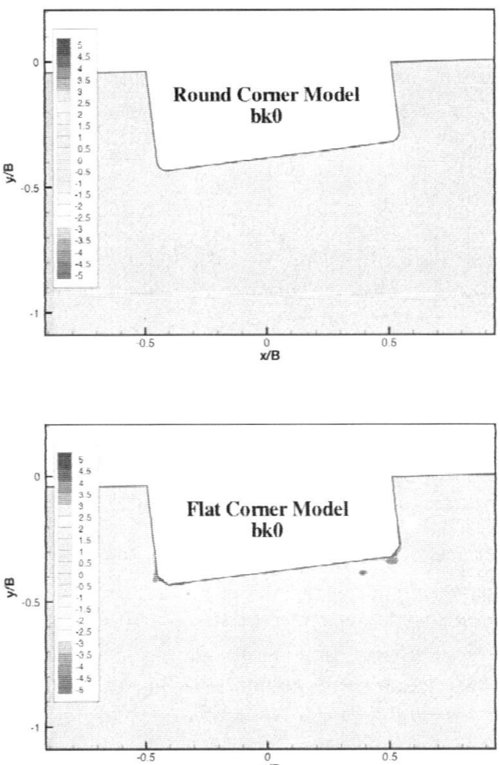

Fig. 13 Vorticity magnitude contours (bk0; t^*=47.7; T_w^*=5.65)

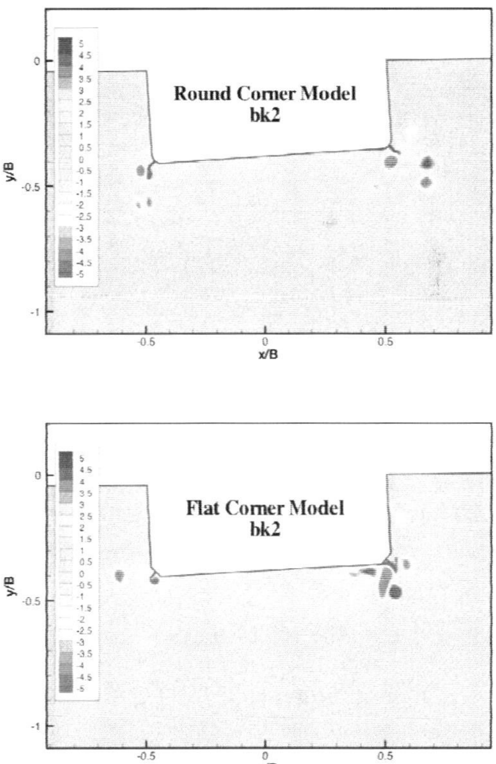

Fig. 14 Vorticity magnitude contours (bk2; t^*=47.7; T_w^*=5.65)

that for the flat corner hull since the bilge keel induced shed vortices are located further away from the hull for the round corner model. In addition, the performance of bilge keels on roll reduction drops as T_w increases when T_w is larger than T_n, and the installation of bilge keels may excite the roll motion for all wave periods larger than a certain value, which was also shown in the study of Esperança et al. (2008). The actual value depends on the wave frequencies and the natural roll frequency of the hull, which is determined by the hull geometries and properties. According to the study of Jung (2004), shed vortices around the hull section seem to excite, instead of reducing, the roll motion when T_w is larger than T_n.

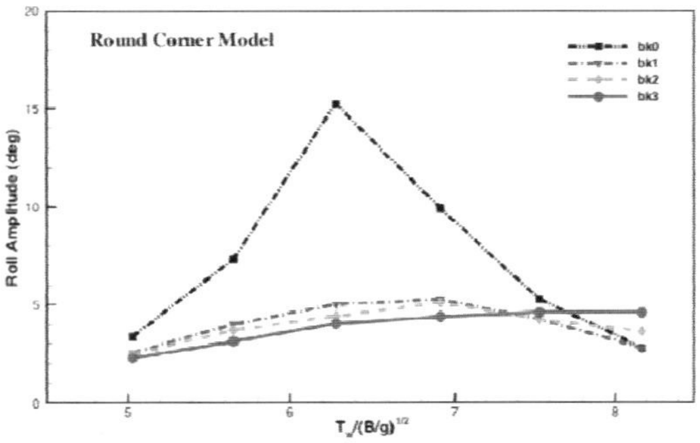

Fig. 15 Maximum roll amplitude for hulls with different bilge keel widths in waves (round corner model)

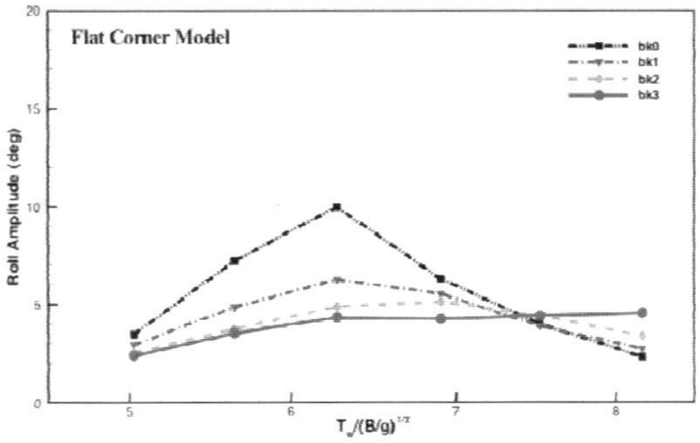

Fig. 16 Maximum roll amplitude for hulls with different bilge keel widths in waves (flat corner model)

CONCLUSIONS

A 2-D Navier-Stokes solver (NS-2D) was applied to model the flow around hull sections (with or without bilge keels) and to predict the corresponding roll response of hulls in regular waves for various incoming wave frequencies, particularly close to resonance. The influence of the hull corner geometry, the bilge keel width and the incoming wave period on roll reduction is studied. The maximum roll response of a hull decreases when a bilge keel is installed, and the reduction increases as the keel width increases. Roll damping is dominated by the location and the strength of the shed vortices. The efficiency of using bilge keels on roll damping therefore depends on the hull geometry, which affects the location of those shed vortices and the ability for waves to carry energy away from the hull in order to reduce the roll motion. In general, the bilge keels are effective on roll reduction over a wide range of wave frequencies, except when the wave period is larger than the hull natural roll-decay period.

ACKNOWLEDGEMENTS

This research was supported by the Offshore Technology Research Center (OTRC), which provides financial support through their Cooperative Agreement with the Minerals Management Service (MMS) and its Industry Consortium.

REFERENCES

[1] DOWNIE, M. J., BEARMAN, P., and GRAHAM, J. "Effect of vortex shedding on the coupled roll response of bodies in waves." *Journal of Fluid Mechanics*, 189 (1988): 243-264.
[2] ESPERANÇA, P. T. T., JR., J. S. S., LIAPIS, S., and MATSUURA, J. P. J. "An Experimental Investigation of Roll Motions of An FPSO." *Proc 27th Int Conf on Offshore Mechanics and Arctic Eng*, OMAE, Estoril, Portugal, (2008).
[3] JUNG, K., CHANG, K., and HUANG, E. "Two-dimensional flow characteristics of wave interactions with a free-rolling rectangular structure." *Ocean Engineering*, 32:1 (2005): 1-20.
[4] JUNG, K. H. "Experimental study on rectangular barge in beam sea Experimental study on rectangular barge in beam sea." PhD Thesis, Department of Civil Engineering, Texas A&M University, College Station, TX, 2004.
[5] KINNAS, S. A., YU, Y.-H., and VINAYAN, V. "Numerical methods for the prediction of the bilge keel effects on the response of ship-shaped hulls." *Proc 26th Int Conf on Offshore Mechanics and Arctic Eng*, OMAE, San Diego, CA, United States, (2007): 1023-1030.
[6] NA, J. H., LEE, W. C., SHIN, H. S., and PARK, I. K. "A Design of Bilge Keels for Harsh

Environment FPSOs." *Proc 12th Int Offshore and Polar Eng Conf*, ISOPE, Kitakyushu, Japan, (2002): 114-117.

[7] RODDIER, D. "Hydrodynamics of Rolling Cylinders With and Without Bilge Keels Hydrodynamics of Rolling Cylinders With and Without Bilge Keels." PhD Thesis, Department of Naval Architecture and Offshore Engineering, University of California at Berkeley, 2000.

[8] VUGTS. "The Hydrodynamic Coefficients for Swaying, Heaving and Rolling Cylinders in a Free Surface." Rept. No. 194, Laboratorium voor Scheepsboukunde, Technische Hogeschool Delft, The Netherlands, 1968.

[9] WILSON, R. V., CARRICA, P. M., and STERN, F. "Unsteady RANS method for ship motions with application to roll for a surface combatant." *Computers and Fluids*, 35:5 (2006): 501-524.

[10] YEUNG, R. W., LIAO, S. W., and RODDIER, D. "Hydrodynamic coefficients of rolling rectangular cylinders." *International Journal of Offshore and Polar Engineering*, 8:4 (1998): 241-250.

[11] YEUNG, R. W., RODDIER, D. R., ALESSANDRINI, L., GENTAZ, L., and LIAO, S. W. "On Roll Hydrodynamics of Cylinders Fitted with Bilge Keels." *Proc 23rd ONR Symp on Naval Hydrodyn*, Val de Reuil, France, (2000): 863-880.

[12] YU, Y.-H. "Prediction of Flows around Ship-shaped Hull Sections in Roll Using an Unsteady Navier-Stokes Solver (also UT-OE Report 08-1) " PhD Thesis, Ocean Engineering Group, Department of Civil, Architectural, and Environmental Engineering, The University of Texas at Austin, Austin, TX, 2008.

[13] YU, Y.-H., and KINNAS, S. A. "Prediction of Hydrodynamic Performance for Various Ship-Shaped Hulls under Excessive Roll Motions Using an Unsteady Navier-Stokes Solver." *Proc 18th Int Offshore and Polar Eng Conf*, ISOPE, Vancouver, BC, Canada, (2008): 386-393.

[14] YU, Y.-H., and KINNAS, S. A. "Roll Response of Ship-Shaped Hulls in Waves." *Proc 28th Int Conf Offshore and Polar Eng*, OMAE, Honolulu, HI, United States, (2009a).

[15] YU, Y.-H., and KINNAS, S. A. "Roll Response of Various Hull Sectional Shapes Using a Navier-Stokes Solver." *International Journal of Offshore and Polar Engineering*, 19:1 (2009b): 46-51.

[16] YU, Y.-H., KINNAS, S. A., VINAYAN, V., and KACHAM, B. K. "Modeling of flow around FPSO hull sections subject to roll motions: Effect of the separated flow around bilge keels." *Proc 15th Int Offshore and Polar Eng Conf*, ISOPE, Seoul, South Korea, (2005): 163-170.

[17] YUCK, R. H., LEE, D. H., CHOI, H. S., JIN, Y. M., and BANG, C. S. "A study on roll damping of 2-D cylinders." *International Journal of Offshore and Polar Engineering*, 13:3 (2003): 205-208.

(This page intentionally left blank.)

PROGRAM SPONSORS

(This page intentionally left blank.)

Wellhead to Market
Your Complete Offshore Solution
INTECSEA is the largest assembly of deepwater expertise in an independent engineering and project management company. INTECSEA offers global full service solutions for pipelines, subsea and floating productions systems.
Visit us at the
Subsea Tieback Conference
Booth 1415
INTECSEA
WorleyParsons Group
intecsea.com
100,000+
km of pipeline
600+
subsea trees
100+
floating systems

Alan C.
McCLURE ASSOCIATES, INC.
Naval Architects • Engineers
1975
2010
acma-inc.com / 713.789.1840

TEXAS SNAME

The Texas Section
of the
Society of
Naval Architects
and
Marine Engineers
would like to thank
our sponsors
for generously
underwriting this
meeting

THE SOCIETY OF NAVAL ARCHITECTS AND MARINE ENGINEERS